An introduction to catastrophe theory

An introduction to

CATASTROPHE THEORY

P. T. SAUNDERS

Lecturer in Mathematics, Queen Elizabeth College
University of London

CAMBRIDGE UNIVERSITY PRESS
Cambridge
London New York New Rochelle
Melbourne Sydney

Published by the Press Syndicate of the University of Cambridge
The Pitt Building, Trumpington Street, Cambridge CB2 1RP
32 East 57th Street, New York, NY 10022, USA
296 Beaconsfield Parade, Middle Park, Melbourne 3206, Australia

First published 1980

Printed in the United States of America
Photoset in Malta by Interprint Limited
Printed and bound by Vail-Ballon Press, Inc.,
Binghamton, New York

Library of Congress Cataloguing in Publication Data
Saunders, Peter Timothy, 1939–
An introduction to catastrophe theory.

Bibliography: p.
Includes index.
1. Catastrophes (Mathematics) I. Title.
QA614.58.S28   514′.7   79-54172
ISBN 0 521 23042 X   hard covers
ISBN 0 521 29782 6   paperback

To the memory of Joseph and Sophie Carpin

Contents

Preface

Almost every scientist has heard of catastrophe theory and knows that there has been a considerable amount of controversy surrounding it. Yet comparatively few know anything more about it than they may have read in some article written for the general public. The aim of this book is to make it possible for anyone with only a modest background in mathematics – no more than is usually included in a first year university course for students not specializing in the subject – to understand the theory well enough to follow the arguments in papers in which it is used and, if the occasion arises, to use it himself.

Most readers will find a number of concepts which are new to them; it would have been impossible to avoid this altogether and still give an adequate account of the theory. But wherever possible I have tried to keep to familiar ground. My object is to explain the theory, rather than to provide formal proofs, and it is almost always harder to understand anything if it is explained in terms of ideas which have themselves only just been introduced. For the same reason, I have sometimes carried out calculations by brute strength and awkwardness when a more elegant derivation was available. I have, however, tried to keep to the spirit, if not always the letter, of the mathematics, and the reader who uses this book as an introduction and then goes on to study the theorems in their full rigour should find that there is nothing he has to unlearn.

Over half the book is devoted to applications, and I have described these in much greater detail than is usual in textbooks on applied mathematics. The reason for this is that it is not possible, certainly not yet, for the mathematician applying catastrophe theory to separate his analysis from the original problem. Had I been writing on, say, new techniques for handling partial differential equations, then I might not have felt obliged to say very much about the problems out of which the equations arose, except perhaps to explain why certain equations were of particular interest, or to provide motivation for the reader. The construction of the models and the interpretation of the results in physical terms could have been taken more or less for granted, leaving the mathematical analysis to be considered on its own.

But with catastrophe theory, especially as it is applied in the biological and social sciences, the situation is quite different. Typically we do not have a model, in the usual sense of the word; in many cases if we had such a model we could manage with the traditional techniques. Instead, we are confronted with a complex system which we are unable to analyse in detail. Our aim is to learn as much as we can without the mechanistic model which is beyond our reach. The approach to be employed depends on the nature of the problem and on our own ingenuity in tackling it; there are as yet no standard techniques. So the best thing to do seems to be to provide a relatively large sample of different applications, to demonstrate the range of possibilities.

There is another reason for including so many applications, and that is this: as I have made no attempt to prove the theorems that are the foundations of catastrophe theory, the reader will have to take them on trust. He can, however, feel reasonably secure; all the mathematicians who have taken the trouble to go through the proofs seem agreed that they are correct. The controversy concerns the applications, and by studying a number of these the reader should be able to make up his own mind.

The examples in the non-physical sciences should be of the greatest value in this, because in my view most of the controversy arises from the fact that catastrophe theory is essentially a part not of theoretical physics, as are most branches of applied mathematics, but of theoretical biology. Thom has written that biologists are not used to thinking in theoretical terms, but while I agree with him wholeheartedly in this, I think he should have added that mathematicians are not used to thinking in biological terms either. We are accustomed to theoretical physics, a subject with a well-established and highly successful paradigm. Unfortunately, this paradigm cannot be carried over unaltered into biology. Nor have the biologists developed the theoretical aspects of the subject sufficiently to provide a ready-made framework for us. Consequently anyone who is trying to apply mathematics in biology finds himself having to work within a paradigm which is not only different from the one that most of his fellow mathematicians take so much for granted as to be scarcely aware of its existence, but which is also still in the early stages of development. Under the circumstances, a full-blown paradigm argument was probably bound to erupt sooner or later. And since catastrophe theory originated not just from theoretical biology, but from that part of the subject which is different in essence from theoretical physics, it is hardly surprising that it has become the centre of the storm.

There remain, to be sure, a number of significant problems concerning catastrophe theory and its applications. But to acknowledge that there are still difficulties connected with a theory that is scarcely a decade old and at the same time to see it as a major advance are in no way incompatible. This point was made many years ago by de Morgan, who wrote

> A disposition sometimes appears to reject all that offers any difficulty, or does not give all its conclusions without any trouble in the examination of apparent contradictions. If by this it be meant that nothing should be permanently used, and implicitly trusted, which is not true to the full extent of the assertion made, I, for one, should offer no opposition to such a rational course. But if it be implied that nothing should be produced to the student, with or without warning, which cannot be understood in all its generality, I should, with deference, protest against a restriction which would tend, in my opinion, not only to give false views of what is already known, but to stop the progress of discovery. It is not true, out of geometry, that the mathematical sciences are, *in all their parts*, those models of finished accuracy which many suppose. The extreme boundaries of *analysis* have always been as imperfectly understood as the tract beyond the boundaries was absolutely unknown. But the way to enlarge the settled country has not been by keeping within it, but by making voyages of discovery, and I am perfectly convinced that the *student* should be exercised in this manner; that is, he should be taught how to examine the boundary, as well as to cultivate the interior. I have therefore never scrupled, in the latter part of this book, to use methods which I will not call doubtful, because they are presented as unfinished, and because the doubt is that of an expectant learner, not of an unsatisfied critic. Experience has often shown that the defective conclusion has been rendered intelligible and rigourous by persevering thought, but who can give it to conclusions which are never allowed to come before him?

The subject being defended in this passage is in fact the calculus, which had been invented about a century and a half earlier but had still not been put on a completely firm foundation; de Morgan was writing only twenty years after the publication of Cauchy's lectures, which contain the first adequate definition of a limit. In the meantime there had been many strong attacks on the calculus. The best known of these is Bishop Berkley's (1734) *The Analyst, or, A Discourse Addressed to an Infidel Mathematician,* and the reader might find it instructive to have a look at this work, if only to see that controversy and polemics are by no means new in mathematics. It is very easy to imagine that mathematics is such a logical subject, and mathematicians such rational people, that each major advance must have been greeted with immediate acceptance and acclaim by the mathematical community, and that consequently anything new which does not achieve this must be without value. Even a cursory glance at the history of our subject shows that this is simply not so.

Of course some of the objections raised by the opponents of the calculus were valid. There was indeed much to be done before the usefulness of the calculus was underpinned by the rigour of analysis. But imagine the loss if the effect of the attacks had been either that the calculus was abandoned altogether, or at least that the eighteenth century mathematicians such as Euler, d'Alembert, the Bernoullis, Lagrange and many others, had refused to work on the applications of the theory until it had been put on a basis as sound as that of geometry.

•Many people have contributed in one way or another to this book. In particular I should like to thank Michael Bazin, Mae Wan Ho and Alan Pears, who suggested improvements to me, though my occasional stubbornness in these matters frees them from responsibility for any errors that remain. I am also grateful to Christopher Zeeman, through whose work I first became acquainted with catastrophe theory, and above all to René Thom. If this book encourages others to take advantage of Thom's great contribution and to build on it, then it will have served its purpose.

1

Introduction

A great many of the most interesting phenomena in nature involve discontinuities. These may be in time, like the breaking of a wave, the division of a cell or the collapse of a bridge, or they may be spatial, like the boundary of an object or the frontier between two kinds of tissue. Yet the vast majority of the techniques available to the applied mathematician have been designed for the quantitative study of continuous behaviour. These methods, based primarily on the calculus, though very much refined and extended since the time of Newton and Leibniz, have made possible tremendous advances in our understanding of nature. Their great success has, however, been largely confined to the physical sciences. When we turn to the biological and social sciences, we generally find that we are unable to construct the relatively complete models which would permit us to apply the same methods. Moreover, the observations, which are the raw material from which the theoretician must work and the standard against which he must test his models, are seldom of the same precision as those which are available in physics. In many cases they are only qualitative. There is nothing in biology to compare with the inexorable and accurately predictable motions of the heavenly bodies.

As a part of mathematics, catastrophe theory is a theory about singularities. When applied to scientific problems, therefore, it deals with the properties of discontinuities directly, without reference to any specific underlying mechanism. This makes it especially appropriate for the study of systems whose inner workings are not known, and for situations in which the only reliable observations are of the discontinuities. It is of course true that the techniques of mathematical physics have been successfully applied to the analysis of discontinuities, but they require a degree of knowledge of the system which workers in the 'soft' sciences are unlikely to possess: the classical treatment of shock waves depends on a detailed understanding of the continuous behaviour of fluids.

Catastrophe theory

Consider a system whose behaviour is usually smooth but which sometimes (or in some places) exhibits discontinuities. We may suppose without much loss that the state of the system at any time can be completely specified by giving the values of n variables $(x_1, x_2, \ldots, x_n)$, where n is finite but may be very large. In a model of the brain, n might be of the order of millions, or hundreds of millions. We may also suppose that the system is under the control of m independent variables $(u_1, u_2, \ldots, u_m)$, i.e. that the values of these variables determine those of the x_i, though not quite (as we shall see) uniquely. We shall suppose that m is relatively small, usually no greater than 5, although we can cope with larger values. This is less of a restriction than it might seem because we are leaving out of account any independent variables that do not have a significant effect on the discontinuity we are studying, and if a system is behaving in a discontinuous fashion and if half a dozen or more independent variables are critically involved, then it is obviously going to be very difficult to make sense of it by any means. We shall refer to the x_i as *state* or *internal* variables, and to the u_i as *control* or *external* variables.

We shall also suppose that the dynamic of the system is derivable from a smooth potential, though in fact the requirement that there be a potential is very much stronger than we really need. It is, for example, sufficient that there be a Liapounov function (which resembles a classical potential in that its minima determine the stable equilibria, but differs from it in that its gradient does not determine the trajectories) so the theory applies to systems which are almost always at equilibria of sets of ordinary differential equations, whether the dynamic is of gradient type or not. Catastrophe theory can also be shown to apply to systems which depend on a variational principle or are governed by many of the commonly encountered partial differential equations, and even in situations in which there are limit cycles, rather than point equilibria. We shall find it convenient in the next few chapters to think in terms of potentials, but this should not be taken to imply that the results are applicable only to systems with gradient dynamics.

The requirement that the potential (more generally, the underlying dynamic) must be smooth is necessary for two related reasons. First, we are interested in the origin of discontinuities, and we have not progressed very far if we simply assume that they are built into the dynamic; this implies that we have not taken our analysis deep enough. And of course, if we allow *a priori* discontinuities in the dynamic we can hardly expect to say very much about their properties, since we will be at liberty to arrange whatever discontinuities we like.

Given these conditions, what catastrophe theory tells us is the following: The number of qualitatively different configurations of discontinuities that can occur depends not on the number of state variables, which may be very large, but on the number of control variables, which is generally small. In particular, if the number of control variables is not greater than four, then there are only seven distinct types of *catastrophes*, and in none of these are more than two state variables involved. (By this last we mean of course that it is possible to choose a set of n state variables in such a way that no more than two are involved in the discontinuity; we may think of this as a transformation to eigen-variables.)

This is a remarkable result. Imagine trying to analyse a large and complex system by conventional means. We would have to write down n differential equations (remember that n may be 10^6 or more), supply initial conditions, solve, and then try to comprehend the solutions. Even if we knew in advance which variables were of interest to us we would be little further ahead. Coupled differential equations cannot be dealt with individually; we would have to solve all n equations first and pick out the one or two relevant solutions afterwards. Catastrophe theory, on the other hand, can make it possible to predict much of the qualitative behaviour of the system without even knowing what the differential equations are, much less solving them. And it does this on the basis of a very few assumptions which are surprisingly unrestrictive.

The proof of this assertion is, as one might expect, difficult, and we are not going to attempt it in this book. The next three chapters will, however, be devoted to an introduction to some of the ideas that lie behind catastrophe theory, and to showing how the list of seven elementary catastrophes arises, and what properties each of the seven has. We shall then go on to discuss examples of the many ways in which the theory can be applied.

Before embarking on this project, we consider in detail two simple physical systems which illustrate a number of the basic themes of catastrophe theory. In particular, they show how it is that a smooth dynamic can give rise to discontinuous behaviour.

The Zeeman catastrophe machine

Fig. 1.1 is a sketch of an educational toy invented by E. C. Zeeman (1972*a*). It is very easy to build, and the reader is strongly advised to make one for himself. The simplest procedure is to find two nearly identical rubber bands. Take the unstretched length of these bands to be one unit. Cut from thin cardboard a disc whose diameter is one unit, and push a drawing pin through the disc at a point Q near the circumference. With the point of this pin upwards, mount the disc on a suitable base by pushing a second pin through the centre, O. Loop the two rubber bands over the pin at Q, and use a third drawing pin to fasten the other end of one of the rubber bands to a point R on the base, two units from O. The remaining end, P, is left free. The dimensions are all approximate; there is no need to adhere to them more than roughly.

We operate the machine by moving P slowly in the plane of the machine. If we experiment for some time, we discover some curious features. The most obvious of these is that while the machine almost

Fig. 1.1. The Zeeman catastrophe machine.

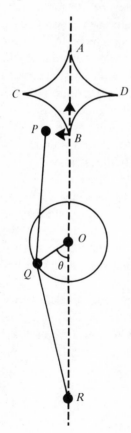

always responds smoothly to small changes in the position of P, it occasionally jumps suddenly. If we mark on the base the positions of P at which these jumps occur, we find that they form the perimeter of a curved diamond. It is, however, possible for P to cross the perimeter without causing a jump. For example, if we move P back and forth across the diamond at right angles to the axis of symmetry of the machine, then there is only one jump in either direction, and these do not occur at the same place. Finally, if P is outside the diamond then only one equilibrium position of the disc is possible, but if P is inside the diamond then there are two stable positions in which it can remain, one with Q inclined to the left, and one with Q to the right. If we are careful we can find a third equilibrium between these two, but it is unstable.

In principle, the analysis of the 'catastrophe machine' is perfectly straightforward. The state of the machine at any time is completely specified by the single variable θ, the angle that the line OQ makes with the axis of symmetry. For any imposed position of the free end P, the machine seeks a configuration (i.e. a value of θ) which minimizes the energy stored in the rubber bands. Since the energy of a stretched elastic is proportional to the square of the extension (the difference between the stretched and unstretched lengths), the potential energy of the system is

$$V(\theta) = \tfrac{1}{2}\mu[(r_1 - 1)^2 + (r_2 - 1)^2],$$

where r_1 and r_2 are the lengths of the rubber bands and μ is their modulus of elasticity.

Because this apparently simple expression turns out to be rather awkward to work with, it is best to begin by considering the special case in which P is allowed to move only along the axis of symmetry. Let the distance OP be s; then (cf. Fig. 1.2)

$$r_1^2 = s^2 + \tfrac{1}{4} + s\cos\theta,$$
$$r_2^2 = 4 + \tfrac{1}{4} - 2\cos\theta.$$

It is still difficult to obtain results in closed form, so we proceed in a different way. By symmetry, there must be an equilibrium position at $\theta = 0$. We can determine the nature of this equilibrium by expanding V as

Fig. 1.2.

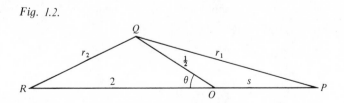

a series in θ, up to terms of order four for reasons which will become clear shortly. To this order

$$r_1^2 \sim s^2 + \tfrac{1}{4} + s(1 - \tfrac{1}{2}\theta^2 + \tfrac{1}{24}\theta^4)$$

and

$$r_2^2 \sim \tfrac{17}{4} - 2(1 - \tfrac{1}{2}\theta^2 + \tfrac{1}{24}\theta^4),$$

whence

$$r_1 \sim (s + \tfrac{1}{2})\left[1 - \frac{s\theta^2}{4(s + \tfrac{1}{2})^2} + \tfrac{1}{16}\left(\frac{s}{3(s + \tfrac{1}{2})^2} - \frac{s^2}{2(s + \tfrac{1}{2})^4}\right)\theta^4\right]$$

and

$$r_2 \sim \tfrac{3}{2} + \tfrac{1}{3}\theta^2 - \tfrac{7}{108}\theta^4,$$

so that

$$V(\theta) \sim \tfrac{1}{2}\mu\left\{(s - \tfrac{1}{2})^2 + \tfrac{1}{4} + \left[\frac{1}{3} - \frac{s(2s - 1)}{2(2s + 1)}\right]\theta^2\right.$$
$$\left. + \left[\frac{s}{24} - \frac{s}{12(2s + 1)} + \frac{s^2}{2(2s + 1)^3} + \frac{5}{108}\right]\theta^4\right\}.$$

The fact that there is no linear term in the expansion tells us that $dV/d\theta$ vanishes at $\theta = 0$, which confirms our assertion that there is always an equilibrium with OQ along the axis of the machine. The nature of this equilibrium is determined by the sign of the second derivative $d^2V/d\theta^2$, i.e. by the sign of

$$\frac{1}{3} - \frac{s(2s - 1)}{2(2s + 1)}.$$

In particular, the equilibrium will change from stable to unstable (or vice versa) at those values of s for which this expression vanishes, i.e. for

$$s = (7 \pm \sqrt{97})/12.$$

The negative root is extraneous, because it corresponds to a position of P for which the elastics are not both stretched. The positive root, however, fixes the point B at $s \approx 1.40$. A similar calculation based on $\pi - \theta$ enables us to locate A at $s \approx 2.46$.

We can now explain at least part of the behaviour of the catastrophe machine. For θ small, the potential is of the form

$$V(\theta) = a + b\theta^2 + c\theta^4.$$

The first two derivatives are accordingly

$$V' = 2b\theta + 4c\theta^3$$

and

$$V'' = 2b + 12c\theta^2.$$

If P is just below B, then b and c are both positive, and so the only real root of the equation $V' = 0$ is $\theta = 0$. There is thus only one equilibrium possible, that with OQ along the axis, and since $V''(0) = b > 0$ the equilibrium is stable.

If we now move P to B, we find that b vanishes. Again the only real root of $V' = 0$ is $\theta = 0$ (note that this is now a triple root) but this time $V''(0)$ vanishes, as indeed does $V'''(0)$. The fourth derivative is positive however, so the equilibrium is still stable; it was to be able to determine this that we carried our expansion as far as the fourth order.

Finally, when P is above B, the coefficient b becomes negative. As a result, the equation $V' = 0$ has three distinct real roots, $\theta = 0$ and $\theta = \pm\sqrt{(-b/2c)}$. We can readily check that the equilibrium corresponding to the first of these is unstable and that the other two are stable.

It follows that if P is moved along the axis from a point outside the diamond to one within it, then Q will remain on the axis until P reaches A or B, when it will move to one side or the other. Once P is inside the diamond there are two equilibrium positions in which the machine will remain if placed, whereas outside there is only one.

There are two things worth noting here. First, A and B have been located exactly, even though we found them by using a Taylor expansion. This is important because the next step will be to expand V about B. Secondly, in finding the equilibria near $\theta = 0$ we assumed that we could ignore all terms in the expansion of V which were of higher order than the first one which was bounded away from zero. This is a standard technique, and it is in fact correct, but it is by no means obvious that we are entitled to do this. One of the lesser known accomplishments of catastrophe theory is that it provides a rigorous justification for the truncation. Not only does this put things on a more secure foundation, but it also makes it comparatively simple to decide what may or may not be neglected in more complicated situations in which our intuition is less reliable.

Having located the cusp points, we can analyse the behaviour of the machine when P is near them. We let B be the origin and take coordinates ξ, η as shown in Fig. 1.3. The expressions for r_2 and for V in terms of r_1 and r_2 are as before, but r_1 is now given by

$$r_1^2 = (s + \tfrac{1}{2}\cos\theta)^2 + (\tfrac{1}{2}\sin\theta - \eta)^2,$$

where

$$s = \hat{s} + \xi, \quad \hat{s} = OB.$$

This gives, to fourth order in θ and to first order in ξ and η,

$$r_1 \sim (s+\tfrac{1}{2}) - (s+\tfrac{1}{2})^{-1}(\tfrac{1}{2}\eta\theta + \tfrac{1}{4}s\theta^2 - \tfrac{1}{12}\eta\theta^3 - \tfrac{1}{48}s\theta^4)$$
$$- \tfrac{1}{8}(s+\tfrac{1}{2})^{-3}(\tfrac{1}{4}s^2\theta^4 + \hat{s}\eta\theta^3).$$

We remark that in both r_1 and r_2, η is a factor of the coefficients of the odd powers of θ but not of the even ones, whereas ξ occurs only in the coefficients of the even powers. We know from the previous calculation that when $\xi = \eta = 0$ the coefficient of θ^2 vanishes but that of θ^4 does not, and it follows that if in each coefficient we retain only the lowest order term in ξ and η then, to fourth order in θ,

$$V(\theta) \sim \tfrac{1}{2}\mu(a_0 + a_1\eta\theta + a_2\xi\theta^2 + a_3\eta\theta^3 + a_4\theta^4).$$

The reader may prefer to verify this by explicit calculation, in which case he should find that the values of the constants a_i are approximately 0.54, -0.24, 0.16, 0.09, 0.45 (Poston & Stewart, 1978a, though attributed by them to Zeeman). In fact the values are not needed for the present discussion.

We can simplify our work by some changes of variable. First, we choose units of elasticity such that $\tfrac{1}{2}\mu a_4$ is equal to unity, which carries V to the form

$$V(\theta) \sim b_0 + b_1\eta\theta + b_2\xi\theta^2 + b_3\eta\theta^3 + \theta^4.$$

We eliminate the cubic term by the substitution

$$x = \theta + \tfrac{1}{4}b_3\eta$$

and we note that to first order in ξ and η this has no other effect on V. Then we define new variables u and v by

$$u = b_2\xi, \quad v = b_1\eta.$$

Finally, since we are interested only in the critical points of V and not in its value, we can without loss change the origin of V to eliminate the constant term b_0. This leaves

$$V(x) \sim x^4 + ux^2 + vx$$

Fig. *1.3.*

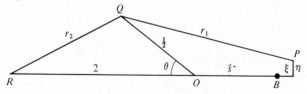

as the simplified form of V with which we shall work; later on this will become familiar as the canonical form of the cusp catastrophe.

The critical points of this function are the solutions of $V' = 0$, i.e. of

$$4x^3 + 2ux + v = 0,$$

and because this is a cubic equation it has either one or three real roots. The number of real roots is determined by the discriminant

$$\Delta = 8u^3 + 27v^2.$$

If $\Delta \leq 0$ there are three real roots; otherwise there is only one. The roots are distinct unless $\Delta = 0$, in which case either two coincide (if u and v are non-zero) or all three coincide (if u and v vanish).

We illustrate this by a diagram of the u–v plane (Fig. 1.4) in which the curve $V(x)$ has been sketched for different values of the parameters u and v, i.e. for different positions of P, the free end of the elastic. We note that

Fig. 1.4. $V(x)$ for different values of u and v.

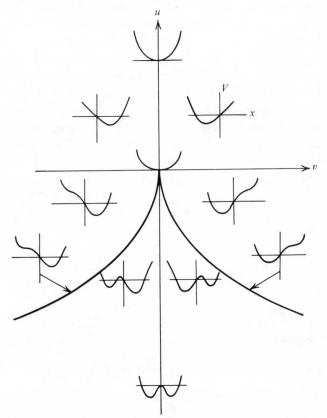

the case of two coincident roots corresponds to a point of inflexion, as a maximum and a minimum coalesce, and we can easily verify that the condition $V' = \Delta = 0$ is exactly equivalent to the condition $V' = V'' = 0$.

Because of the simple relation between the variables (x, u, v) and the variables (θ, ξ, η), these sketches also represent the form of the curve $V(\theta)$ for different values of ξ and η. As a result, if we restrict ourselves to the region near the point B, we can now account for the behaviour of the machine in the more general case in which P is no longer constrained to move along the axis. So long as P is outside the cusp, there is only one equilibrium possible; when it is inside there are two. Jumps occur on leaving the cusp, but only if the equilibrium that disappears is the one in which the machine happens to be. This is why leaving the cusp by the same path along which it was entered does not produce a jump.

Fig. 1.5 represents a set of potentials for different values of v and a fixed negative value of u. It may be useful to think of it in a mechanical analogy, with the curve being made of a flexible material and the system being represented by a ball bearing. One possible sequence of events is illustrated in the figure; the reader can readily construct others. Note that the analogy depends on there being enough friction to prevent the bearing shooting through a local minimum and out the other side. This serves to remind us that there is generally an equilibrium or 'quasi-stationary' assumption involved when we use elementary catastrophe theory; it does not necessarily apply to essentially dynamic situations.

It is easier to predict what will happen for different paths in the neighbourhood of B if we sketch the surface $4x^3 + 2ux + v = 0$ (Fig. 1.6). This is the set of equilibrium values of (x, u, v) for the system. Hence if we think of the state of the system as being represented by a point in the three-dimensional *phase space* with x, u, v as coordinates, the *phase point* must always lie on the surface. In fact it must always lie on either the top or the bottom sheet, because the middle sheet corresponds to unstable equilibria.

We interpret the diagram as follows. The location of the point P is represented by a point in the u–v plane, which we call the *control space*.

Fig. 1.5. $V(x)$ for fixed $u < 0$ and different values of v. After Zeeman (1976a).

As the control variables u and v are altered, this control point traces out a path which we call the *control trajectory*. At the same time, the phase point moves along a trajectory in the equilibrium surface, directly above the control trajectory. Smooth variations in u and v almost always produce smooth variation in x. The only exceptions occur when the control trajectory crosses the *bifurcation set* $8u^3 + 27v^2 = 0$, which is the projection onto the u–v plane of the folds of the equilibrium surface. If the phase point happens to be on the surface which ends at this point (by folding back as the middle sheet) then it must jump to the other sheet. This brings about a sudden change in x, i.e. in θ.

This picture is also useful when we are dealing with more complex systems. If there are n state variables and m control variables, then the phase space will be an $n+m$ dimensional Euclidean space, which we denote by R^{n+m}. As the control point moves in the m-dimensional control space, the phase point moves 'above' it on the equilibrium hypersurface, and sudden jumps occur when the control trajectory crosses the bifurcation set, which is the projection into the control space

Fig. 1.6. Phase space for the Zeeman catastrophe machine, showing the equilibrium surface and the control space. Also shown are the trajectory and control trajectory corresponding to the sequence illustrated in Fig. 1.5.

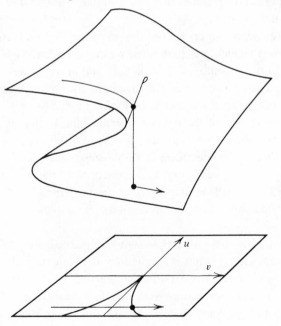

of the folds of the equilibrium hypersurface. This may sound somewhat forbidding, but in fact R^{n+m} is in many ways very much like R^3, and the reader should find that he can understand what happens in the higher dimensional cases quite adequately by thinking of them in terms of Fig. 1.6.

Zeeman (1976a) has suggested five properties which are typical of the cusp catastrophe. We shall want to make use of the list later and we can now demonstrate these properties by means of Fig. 1.6, or on a real catastrophe machine or (preferably) both.

The first is the occurrence of *sudden jumps*, which we have already seen. The second is *hysteresis*: if we move P back and forth across the curved diamond, the jumps to the right do not occur at the same place as the jumps to the left. The third is *divergence*: if we start with P on or near the axis but outside the diamond and move it to a point inside, then whether Q is to the right or to the left depends only on which side of B the free end P passed, so two nearby trajectories connecting the same two points can produce significantly different behaviour. Related to this is *bimodality*: for certain positions of P (those within the diamond) there are two possible stable positions of Q. Finally, there is *inaccessibility*: if P is above or below the diamond we can give θ any value we like by moving P along a line at right angles to the axis, but if P is level with the diamond there are certain values of θ (for example, $\theta = 0$) at which stable equilibrium is not possible.

All these are easily observed on the machine. The advantage of pointing them out explicitly is that when we come to deal with systems whose mechanisms we cannot analyse directly and in which we may only suspect that the cusp catastrophe is involved, the list serves to remind us of the sorts of phenomena we should be looking for.

The chief importance of the catastrophe machine is that it demonstrates how a smooth potential can give rise to discontinuous behaviour, through the disappearance of stable steady states. This idea is fundamental to catastrophe theory, and consequently the mathematical arguments which we shall be discussing in the next three chapters will be primarily concerned with configurations of stable and unstable equilibria.

There are, however, some other lessons to be learned. First of all, we can now see more clearly what is implied by the statement that catastrophe theory yields qualitative results. We carried out the analysis by standard methods, and our results were therefore quantitative. Yet this was not really what we wanted; it was at the same time more and less than was required. Our primary concern was not to predict the behaviour of the particular machine illustrated in Fig. 1.1, some of whose dimensions were chosen purely for convenience in performing the calculations. What we were actually trying to do was to understand why a machine of that general design exhibits sudden jumps in a rather peculiar way. By using conventional methods we were able to locate precisely the points at which the discontinuities would occur, but only for one somewhat idealized machine. What we did not show – which is what we really wanted to know and which is generally, though not always correctly, taken to be 'obvious' – is that for any roughly similar machine the pattern of jumps would be the same and that the set of points at which they can occur would form a curved diamond. It is this sort of qualitative result which catastrophe theory provides.

There is also the important question of the 'local' nature of the results. Our analysis was indeed local, in that we were working in an infinitesimal neighbourhood of the point B. Yet the results are valid over a much larger region; in fact once we have located all four cusp points the series expansions enable us to predict the qualitative behaviour of the machine completely (Zeeman, 1972a; for a full quantitative analysis see Poston & Woodcock, 1973). This is characteristic of many applications of catastrophe theory as well, and indicates that we are to interpret the word 'local' to mean 'associated with a given singularity'. In the case of the catastrophe machine, catastrophe theory (like the analysis we actually carried out) would predict the pattern of jumps 'near' each of the cusps, i.e. over the half-diamond adjacent to each. Beyond this the local analysis no longer applies, as the system is under the influence of a different *organizing centre*.

A gravitational catastrophe machine

Poston (1976) has devised a number of catastrophe machines whose potential energy is gravitational, and one of these is shown in Fig. 1.7. The recommended method of construction is as follows: cut a parabola out of cardboard, and cut a second piece of card into a parabola-shaped ring whose outer edge is congruent to that of the first piece. Some care must be taken to make the shapes smooth, with no flat parts or wobbles. Make about six struts out of balsa or folded cardboard, and use them to fasten the two pieces together with their boundaries aligned. Place a small heavy magnet behind the solid parabola and a piece of metal (or another magnet) in front.

We want to predict what will happen as the magnet is slid slowly across the face of the machine. We take the axis of the parabola as the x-axis and the vertex as the origin, in the usual way. The equation of the perimeter is then $y^2 = 4ax$, and the coordinates of any point on it may be expressed in parametric form as $(at^2, 2at)$.

Let the magnet be placed at the point (X, Y) and let the point on the edge which is in contact with the table have parameter t. The equation of the tangent at this point is

$$x - ty + at^2 = 0$$

and the perpendicular distance from (X, Y) to this line (i.e. the height of the magnet above the table) is

$$h = \frac{X - tY + at^2}{\sqrt{(1 + t^2)}}.$$

Since the magnet is very much heavier than the rest of the machine, we may take (X, Y) to be the centre of gravity of the system. The potential energy is then mgh, where m is the mass of the magnet and g the gravitational constant. To locate the equilibrium position of the machine we set dh/dt equal to zero:

$$0 = [at^3 + (2a - X)t - Y][1 + t^2]^{-3/2}.$$

Fig. 1.7. A gravitational catastrophe machine.

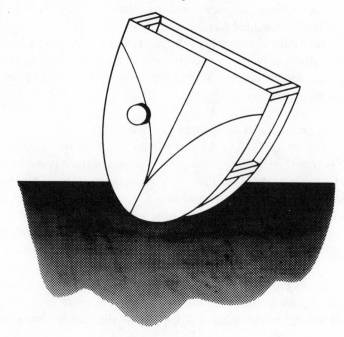

Since $1+t^2$ cannot vanish, the condition for equilibrium is a cubic equation, and we appear to be on familiar ground. In fact we are, but we have to check this, because since t also appears in the other factor it is not obvious how the higher order derivatives will behave.

What actually happens is quite instructive. Consider two functions, $u(x)$ and $v(x)$, and suppose that $u(x_0)=0$ and $v(x_0)>0$. Let $f(x)=uv$. Then, by Leibniz's rule,

$$f'=u'v+uv',$$
$$f''=u''v+2u'v'+uv'',$$
$$f'''=u'''v+3u''v'+3u'v''+v''',$$

and so on. Since $u(x_0)=0$, we have

$$f'(x_0)=u'(x_0)v(x_0)$$

so that at x_0, f' has the same sign as u'. If u' vanishes at x_0, then so too does f', and then f'' and u'' have the same sign. If both u' and u'' vanish at x_0, then so too do both f' and f'', and then f''' has the same sign as u''', and so on.

Thus if the first r derivatives of u vanish, then (and only then) so do the first r derivatives of f and, what is more, the $r+1$th derivatives of the two functions have the same sign. It follows that for determining the number of critical points and their types, we do not have to work with dh/dt itself, but may use instead the simpler function

$$at^3+(2a-X)t-Y.$$

Indeed, since what matters is not the potential itself but only its derivatives, we see that *for our purposes* the machine behaves as though its potential were the polynomial

$$V(t)=\tfrac{1}{4}at^4+(a-\tfrac{1}{2}X)t^2-Yt.$$

By comparing $V(t)$ with the potential we derived for the Zeeman catastrophe machine, we can now predict the behaviour of this machine. We draw on the face of the parabola the cusp-shaped curve whose equation is

$$27ay^2+4(2a-x)^3=0.$$

So long as the magnet is not moved across this curve, the machine will respond to its motion by a smooth change in the point of contact with the table. If, however, the magnet enters the cusp region by crossing one branch of the curve and leaves it by crossing the other, then the machine will shift abruptly to a quite different position.

This is an illustration of an idea which is very important in catastrophe theory. The actual potential energy of the machine is of course *mgh*, and it is this we must compute if we need to know the energy or if we wish to predict the dynamical behaviour of the machine – if, for example, we want to know how quickly it will move to an equilibrium position. But for locating the equilibria and determining their stability, and for predicting where and under what conditions the discontinuities will occur, we may replace the potential by a polynomial with the same critical points. Here we found the polynomial by deriving the true potential first; had we used catastrophe theory we could have deduced almost immediately that the polynomial was a quartic and there would have been no need to work out the potential.

2

Background

Like any other major advance, catastrophe theory depends on a number of concepts with which the reader new to the field is unlikely to be familiar. The aim of this chapter is to provide an introduction to these which is as simple as possible. A more sophisticated, but still relatively accessible, account is given by Poston & Stewart (1978a). Mathematicians might prefer Lu (1976) or, for full proofs, Bröcker & Lander (1975) or Trotman & Zeeman (1976).

Structural stability

Implicit in science is the belief that there is some sort of order in the universe and that, in particular, experiments are generally repeatable. What is often not recognized is that what we demand of nature in this regard is not mere repeatability, but something rather more. It is never possible to reproduce exactly the conditions under which an experiment was performed. The quantity of one of the reagents may have been altered by 0.001%, the temperature may have increased by 0.0002 K, and the distance from the laboratory to the moon will probably be different as well. So what we really expect is not that if we repeat the experiment under precisely the same conditions we will obtain precisely the same results, but rather that if we repeat the experiment under approximately the same conditions we will obtain approximately the same results. This property is known as structural stability. It is not really very different from the sort of stability we are accustomed to in elementary mechanics, except that the system is to be resistant not to disturbance from an equilibrium position but to perturbations of the conditions of the experiment.

The concept of structural stability is also found in mathematics; in fact it originated there. Consider for example, an m-parameter family of functions. If the parameters vary continuously we may think of them as coordinates in an m-dimensional space, and then each individual function is represented by a point in this space. If f_P is the function corresponding to the point P, and if for any Q sufficiently close to P the corresponding function f_Q has the same form as f_P, then f_P is called a structurally stable or *generic* function of the family. The set of all points P whose corresponding functions are generic is called the subset of generic points. The complement of this set, i.e. the set of all points P such that f_P is not generic, is called the *bifurcation set*.

We can also define structural stability for a family (i.e. a sub-family of the one we began with) of functions. In this case we require that any small perturbation leaves the qualitative nature as a family unchanged. This means that whatever forms individual members of the family have must occur in both the original and perturbed families, and that the topological structure of the bifurcation set must be preserved.

There are two points to keep in mind when using these definitions. First, we have only defined structural stability in terms of a given family of functions, and we have left undefined the expression 'of the same form'. In any particular case the definitions must be completed to suit the problem at hand. This may seem odd, but in fact the same is true of ordinary stability: we have to specify for each problem separately what sorts of perturbations we are allowing and which final states are to be considered as equivalent to the initial state.

There is also a slight possibility of confusion arising from the fact that we normally refer to $f(x) = x^4 + ux^2 + vx$ as a function, even when we really mean a two-parameter family of functions. To avoid sounding awkward and pedantic we shall follow the common usage, and we shall not use the expression 'family of functions' except where it is necessary to avoid ambiguity. Thus when we say that the above function $f(x)$ is structurally stable, we shall mean that it is stable if considered as a two-parameter family. Alternatively, we may think of the statement as meaning that $f(x)$ is stable for almost all values of u and v, though only if we adopt the mathematicians' convention that 'almost all' can mean 'with a possibly infinite number of exceptions'. This allows us to say, for example, that almost all the points in the plane do not lie on the x-axis.

As a specific example, let us take as our family of functions the polynomials in one variable x and of degree less than or equal to N. We can take N as large as we like, but by insisting that it be finite we avoid what are for us unnecessary complications. The coefficients may be considered as the parameters of the family, so that two polynomials will be said to be 'close' if all their coefficients are close in the usual sense of the word. With an eye on the potentials we were discussing in the first chapter, we shall say that two polynomials are of the same type if they have the same configuration of critical points near $x=0$.

One member of this family is x^4. To determine whether or not it is stable, we compare it with the neighbouring polynomial

$$W(x)=x^4+\alpha x^p,$$

where $|\alpha|$ is small and p is an integer. Now x^4 has a minimum near the origin, and for $p\geq 4$ so does $W(x)$. If $p=3$, however, then $W(x)$ has a point of inflexion at the origin and a minimum at $x=-\frac{3}{4}\alpha$, while if $p=2$ and $\alpha<0$ then $W(x)$ has a maximum at the origin and minima at $x=\pm\sqrt{(-\frac{1}{2}\alpha)}$. If $p=1$, then $W(x)$ has a minimum at $x=\sqrt[3]{(-\frac{1}{4}\alpha)}$, but it does not follow that the addition of a small linear term cannot alter the type: the function

$$x^4+\alpha x^2+\beta x$$

can (as we have already seen) have a minimum and a point of inflexion near $x=0$, and so is not of the same type as $x^4+\alpha x^2$.

It is important to bear in mind that we are concerned only with what happens arbitrarily close to the origin. Given any $\varepsilon>0$, we can always choose a polynomial of the form $x^4+\alpha x^2$ sufficiently close to x^4 (i.e. $|\alpha|$ sufficiently small) to bring the extra critical points to within ε of $x=0$. This is not true for $x^4+\alpha x^5$ because the additional critical point is at $x=-4/5\alpha$, so that choosing $|\alpha|$ arbitrarily small moves the critical point arbitrarily *far* from the origin.

Thus x^4 is not structurally stable, because there are polynomials close to it which are not of the same type. The family $x^4+\alpha x^2$ is also not structurally stable. On the other hand, the family

$$\tilde{V}(x)=x^4+\alpha x^3+\beta x^2+\gamma x+\delta$$

is structurally stable, since the addition of a fifth or higher order term cannot affect the type, and there are no lower order terms left to add.

We do not have to add all the lower order terms to obtain a structurally stable function. Any polynomial of the family $\tilde{V}(x)$ can be written in a form with no cubic or constant term simply by changing the origin. Consequently all the types encountered in the family $\tilde{V}(x)$ occur in the family

$$V(x)=x^4+ux^2+vx$$

as well, and it follows that the family $V(x)$ is structurally stable. Thus the unstable polynomial x^4 can be stabilized by the addition of two terms.

We call $V(x)$ an *unfolding* of the singularity x^4. The idea is that while x^4 appears to have only one critical point, it is more accurate to say that it has three coincident critical points. By a suitable perturbation of the function we can separate these, and we then discover a much richer range of types than was at first apparent. It is like the unfolding of a bud to reveal a flower.

Structurally stable unfoldings are called *versal*; both $\tilde{V}(x)$ and $V(x)$ are versal, whereas $x^4 + \alpha x^2$ is an example of an unfolding which is not versal. Unfoldings which, like $V(x)$, are versal and have the minimum number of parameters for a versal unfolding are called *universal*. One of the important theorems that we are not going to prove justifies this name by establishing that any two universal unfoldings of the same singularity are equivalent.

Up to this point we have been discussing only polynomials, but this is not as much of a restriction as it might appear, since any sufficiently smooth function of one variable can be expanded – at least formally – in a Taylor series. The coefficient of x^n is (apart from the factor $1/n!$) just the nth order derivative $f^{(n)}(0)$, and this provides us with another way of seeing what is going on. We know from elementary calculus that the nature of a critical point of a function of one variable is usually determined by the sign of the second derivative. If this derivative is zero, however, we have to look at the third derivative; if that too is zero we have to go on to the fourth derivative, and so on. If we have to go on n terms beyond the second derivative before we can tell what sort of critical point we have, we say that the function has an n-fold degeneracy, and we find that it takes n unfolding parameters to stabilize it, one for each missing term.

In almost all cases we eventually find one term in the series which, on account of the nature of the potential, can never vanish. For the systems we were discussing in the first chapter, this was the x^4 term. Once this happens, the rest of the series can be ignored, as this term will determine the nature of the critical point if nothing else does first. An important consequence of this is that we do not have to be concerned about the behaviour of the remainder or whether the series converges (and, if so, to what); as Zeeman has put it, we no longer allow the tail of the series to wag the dog.

The $V(x)$ which we have just found to be the universal unfolding of the singularity x^4 is the same function which we met in our study of the catastrophe machines, and this clue enables us to connect the two kinds of structural stability we have described. The Zeeman machine does not always respond stably; there are circumstances in which a small change in the position of the free end does produce a significant change in the orientation of the pointer. The instabilities are, however, isolated; the machine generally responds smoothly, then undergoes a single abrupt change, then responds smoothly again. Moreover the behaviour is stable in the sense that if we repeat an experiment as closely as we can, we expect the same sort of sequence, with the sudden jump at approximately the same place. We may say that the structural instability occurs in a structurally stable fashion.

The mathematical expression of these ideas is that the process is governed by a family of potentials which is itself structurally stable but which contains a nowhere dense subset of non-generic potentials. (For our purposes, 'nowhere dense' may be taken to mean 'of lower dimension than the family as a whole': this ensures that 'almost all' trajectories in the control space intersect the subset in isolated points.) It is the subset that provides the discontinuities which make the process interesting, while the structural stability of the family ensures the approximate repeatability of experiments.

Not all systems in nature are structurally stable, nor indeed are all dynamical systems in mathematics. In both cases we can find examples of systems with instabilities which are not isolated. We shall not be dealing with such essentially unstable systems in this book. It is easy enough to justify restricting our attention to stable systems, as they are precisely the ones that are amenable to analysis, but this is not to deny the existence of the others.

The splitting lemma

Most systems cannot be adequately described by a single state variable, so before we can apply the ideas we have been developing we have to see what happens in the more general case. We begin by reviewing the analysis of critical points of functions of two variables; not only does this provide us with sufficient insight into the additional sorts of things that can happen, but it also turns out that all that we are going to be doing later will involve only one- or two-variable potentials.

Let $f(x, y)$ be a smooth function of x and y, and suppose that there is a critical point at the origin, so that (using subscripts to denote partial derivatives)

$$f(0, 0) = f_x(0, 0) = f_y(0, 0).$$

Then the Taylor series for f will be

$$f(x, y) = \tfrac{1}{2}(ax^2 + 2hxy + by^2) + \text{higher order terms},$$

where

$$a = \frac{\partial^2 f}{\partial x^2}, \ h = \frac{\partial^2 f}{\partial x \partial y}, \ b = \frac{\partial^2 f}{\partial y^2}.$$

Now it is well known (see almost any text on analytical geometry) that the curve

$$ax^2 + 2hxy + by^2 = F,$$

where F is a constant, is a conic section. If $ab - h^2 > 0$ then it is either an ellipse (if $aF > 0$) or has no real points (if $aF < 0$). If $ab - h^2 < 0$ then it is a hyperbola, with the sign of aF determining which of the two principal axes is the transverse axis. It follows (by considering the intersections of the surface $z = f(x, y)$ with planes $z = \pm \varepsilon$ where $|\varepsilon|$ is small) that sufficient conditions for the following kinds of critical points of the function $f(x, y)$ are

maximum: $\Delta > 0, \quad \partial^2 f / \partial x^2 < 0,$

minimum: $\Delta > 0, \quad \partial^2 f / \partial x^2 > 0,$

saddle: $\quad \Delta < 0.$

For convenience we have written

$$\Delta = \frac{\partial^2 f}{\partial x^2} \frac{\partial^2 f}{\partial y^2} - \left(\frac{\partial^2 f}{\partial x \partial y} \right)^2.$$

The case $\Delta = 0$ remains to be decided.

In the neighbourhood of a non-degenerate critical point, a function of one variable can be closely approximated by a parabola, opening upwards for a minimum or downwards for a maximum. The generalization for a function of two variables is that near a non-degenerate critical point it can be closely approximated either by an elliptic paraboloid (for a maximum or a minimum) or a hyperbolic paraboloid (for a saddle point). See Fig. 2.1.

This is the point at which most calculus texts leave the problem, on the grounds that almost all cases have been covered, but it is also precisely the point at which things start to get interesting, at least for us. Functions for which $\Delta = 0$ are structurally unstable, since there will be functions arbitrarily close to them with $\Delta > 0$ and with $\Delta < 0$, and these will generally be of different types.

There are, however, two quite distinct cases, and we have to distinguish between them. It may be that Δ vanishes because all three second order partial derivatives are zero at the origin, and it is then clear that $f(x, y)$ is degenerate in both the x-direction and the y-direction. On the other hand, suppose that $f_{xx}f_{yy}=(f_{xy})^2$ but that not all the derivatives vanish separately. Then the fact that if $ab-h^2=0$ then $|ax^2+2hxy+by^2|$ is a perfect square allows us to write

$$f(x, y) = \pm \tfrac{1}{2}(px+qy)^2 + \text{higher order terms},$$

where

$$p=\sqrt{|a|}, \qquad q=\sqrt{|b|}.$$

The form of the expansion suggests that we rotate the axes, transforming to new coordinates u, v given by

$$u=\frac{px+qy}{\sqrt{(p^2+q^2)}}, \qquad v=\frac{qx-py}{\sqrt{(p^2+q^2)}}.$$

It is now a straightforward exercise to calculate the first and second order partial derivatives of f with respect to u and v. The values at the origin are

$$\frac{\partial f}{\partial u}=\frac{\partial f}{\partial v}=\frac{\partial^2 f}{\partial v^2}=\frac{\partial^2 f}{\partial u \partial v}=0,$$

Fig. 2.1. Critical points in three dimensions: (*a*) maximum, (*b*) minimum, (*c*) saddle.

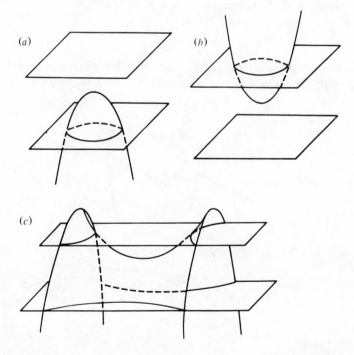

(*a*)

(*b*)

(*c*)

$$\frac{\partial^2 f}{\partial u^2} = \pm(p^2 + q^2) \neq 0.$$

Thus f has either a maximum or a minimum (depending upon the sign) in the u-direction, but we do not yet know what happens in the v-direction. The surface $z = f(x, y)$ is, to second order, a parabolic cylinder (Fig. 2.2).

In fact we know what happens in every direction except the v-direction. For let $w = u \sin\theta + v \cos\theta$. Then at the origin

$$\frac{df}{dw} = \sin\theta \frac{\partial f}{\partial u} + \cos\theta \frac{\partial f}{\partial v} = 0$$

and

$$\frac{d^2 f}{dw^2} = \sin^2\theta \frac{\partial^2 f}{\partial u^2} + 2\sin\theta \cos\theta \frac{\partial^2 f}{\partial u \partial v} + \cos^2\theta \frac{\partial^2 f}{\partial v^2}$$

$$= \sin^2\theta \frac{\partial^2 f}{\partial u^2}.$$

Hence f has the same sort of behaviour in the w-direction as in the u-direction, provided only that $\theta \neq 0$. If θ is equal to zero, i.e. in the v-direction, the Taylor series for f reduces to

$$f = \frac{v^3}{3!} \frac{\partial^3 f}{\partial v^3} + \frac{v^4}{4!} \frac{\partial^4 f}{\partial v^4} + \dots$$

and we are back to the single variable case that we have already discussed. The variable u plays no further role in the analysis. We have

Fig. 2.2.

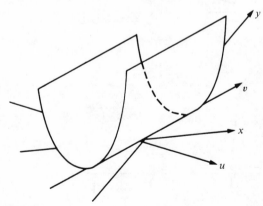

shown, therefore, that if $\Delta = 0$ then the function f is doubly degenerate only if all the second order partial derivatives vanish separately. If they do not, then by a simple coordinate transformation we can reduce the problem to the study of a function of one variable only.

This result can be extended to any finite number of variables. Let $f(x_1, x_2, \ldots, x_n)$ be a function of n independent variables with a critical point at the origin, so that f and all its first order partial derivatives vanish there. Then we form the following matrix, called the *Hessian* of f:

$$\begin{bmatrix} \dfrac{\partial^2 f}{\partial x_1^2} & \dfrac{\partial^2 f}{\partial x_1 \partial x_2} & \dfrac{\partial^2 f}{\partial x_1 \partial x_3} & \cdots & \dfrac{\partial^2 f}{\partial x_1 \partial x_n} \\[2ex] \dfrac{\partial^2 f}{\partial x_2 \partial x_1} & \dfrac{\partial^2 f}{\partial x_2^2} & \dfrac{\partial^2 f}{\partial x_2 \partial x_3} & \cdots & \dfrac{\partial^2 f}{\partial x_2 \partial x_n} \\[2ex] \cdots & \cdots & \cdots & & \cdots \\[2ex] \dfrac{\partial^2 f}{\partial x_n \partial x_1} & \dfrac{\partial^2 f}{\partial x_n \partial x_2} & \dfrac{\partial^2 f}{\partial x_n \partial x_3} & \cdots & \dfrac{\partial^2 f}{\partial x_n^2} \end{bmatrix}.$$

It can be shown that if the rank of the Hessian is n, i.e. if its determinant does not vanish, then there exists a coordinate transformation which permits us to write f in the form

$$f = e_1 x_1^2 + e_2 x_2^2 + \ldots + e_n x_n^2 + \text{higher order terms.}$$

Each of the constants e_i is equal to ± 1. We can read off the type directly from the numbers of plus and minus signs, and f is structurally stable.

If, on the other hand, the rank of the Hessian is $n - r$ for some $r > 0$, then there exists a coordinate transformation which permits us to write f in the form

$$f = e_{r+1} x_{r+1}^2 + e_{r+2} x_{r+2}^2 + \ldots + e_n x_n^2 + \text{higher order terms.}$$

The structural instability is confined to the variables $x_1, x_2, \ldots, x_r$ and can be analysed in terms of these variables alone; the remaining variables $x_{r+1}, x_{r+2}, \ldots, x_n$ can be ignored.

This result is called the 'splitting lemma' because it allows us to split the variables into two classes: 'essential variables' which are involved in the structural instability and 'inessential variables' which are not – and then to ignore the second class. The number of kinds of catastrophes which can occur depends not on the number of state variables, n, but only on the number of essential state variables, r. We call this number the *corank* of the Hessian, and, equally, of the function f. We may think of it as specifying the number of directions in which the function is degenerate.

Codimension

Having learned that a single equation generally represents a curve in the plane, students are often surprised, when they progress to solid analytical geometry, to discover that a single equation now represents not a curve but a surface. And if they continue to four-dimensional geometry, for example if they study relativity theory, they find that a single equation represents a three-dimensional hypersurface, with two equations needed for a 2-surface and three for a curve. By this time, a pattern has emerged: the number of equations required to represent a geometric object is in general equal to the difference between the dimension of the object and that of the space in which it is embedded. We call this quantity the *codimension* of the object.

Objects with the same codimension have certain properties in common, apart from (though not, of course, independent of) requiring the same number of equations. For example, only an object of codimension 1 can divide R^n into two distinct parts. Thus a point divides a line, a line divides a plane (so does a simple closed curve, into an inside and an outside), a plane divides R^3, and the whole of space–time (if we take it to have the same global topology as R^4) is divided into past and future by the three-dimensional 'snapshot' of the universe now, which we may take as the hypersurface $t = t_0$.

Codimension has the important property that it is generally preserved if we change the dimension of a problem by ignoring inessential coordinates. This is illustrated by Fig. 2.3, in which is shown part of R^3 with a curve in it. Suppose we now decide to ignore the z-coordinate, and restrict our attention to the x–y plane. The curve becomes a single point, which is an object with a different dimension (zero, not one) but the same

Fig. 2.3.

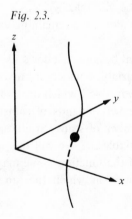

codimension, two. We see from this example that codimension (like corank, with which it is not to be confused) is an appropriate quantity with which to work in a subject in which we are constantly trying to reduce the dimensions of problems to make them more manageable – and in which the actual dimension of the problem (i.e. the number of state variables) may not even be known.

In general, a 1-parameter continuous family of objects of dimension p is an object of dimension $p+1$: a family of points is a curve, a family of lines is a surface, and so on. It is easy to see why this is so; we have only to set up a coordinate system on each member of the family and then any point in the new geometric object will be specified by $p+1$ coordinates: the p coordinates it has on the particular member of the family and the parameter which tells us which member of the family that is. In the same way, an r-parameter continuous family of objects of dimension p is an object of dimension $p+r$. Equally, an r-parameter family of objects of codimension p is an object of codimension $p-r$.

Now consider a point P which lies on an object of codimension r in some Euclidean space R^m. Suppose that we want to construct a family of similar objects such that there is a neighbourhood of P every point of which lies on a member of that family. This can only be done if the geometric object generated by the family is of dimension m, and so we will need an r-parameter family.

We now interpret R^m as the parameter space for polynomials in one variable, x, and of degree not greater than m, and which have a critical point at the origin. The subset of parameter values which correspond to functions f such that $f_{xx}=0$ is a geometric object of codimension 1, since it is specified by one equation. The subset of parameter values for which $f_{xx}=f_{xxx}=0$ is a geometric object of codimension 2, since it is specified by two equations. We can now see on geometric grounds why it is that the structurally unstable functions x^3 and x^4 require one and two unfolding parameters, respectively, and we shall say that the functions themselves have codimension equal to the number of parameters in the universal unfoldings.

Of course for functions of one variable we have no real need of all this analysis. It is reasonably clear in the first place what has to be done: we need all the lower order terms, apart from one which can be got rid of by a change in the origin of x, and the constant, which can be eliminated by a change in the origin of f. Suppose, however, that we have a function of n variables. Then, as we have seen, the condition that the function is structurally unstable and that the problem does not reduce to one in fewer than n variables means that a quadratic form in n variables vanishes identically, i.e. that the rank of the Hessian is zero. If $n = 1$ then this condition is simply $f_{xx} = 0$; this is one equation and so the codimension can be as small as 1. If $n = 2$, however, the condition is $f_{xx} = f_{xy} = f_{yy} = 0$, which is three equations. Hence any function which is degenerate in two directions is of codimension at least 3, and requires at least three unfolding parameters. In general, the minimum codimension for a function of corank n is $\frac{1}{2}n(n+1)$, the number of independent entries in an $n \times n$ symmetric matrix.

This establishes half the result that was promised in the first chapter, because we can now see that a singularity with 3 essential state variables must have codimension at least 6. If, therefore, we restrict our attention to processes with no more than 5 control variables, the number of essential state variables cannot be greater than 2. In the next chapter we shall derive the canonical forms of all the different catastrophes that can arise, and in so doing we shall show that with 4 or fewer control variables there are only 7, while if we allow 5 control variables the number rises to 11. We shall also indicate why it is that once the codimension reaches 6, things become more complicated.

The reader may have noticed that we have reached the end of this chapter without saying much about the 'deep mathematical theorems' which he has probably heard are the foundations of catastrophe theory. The reason for this is twofold. First, not only are the theorems hard to prove, they are even hard to state in terms of the limited mathematical vocabulary we are using in this book. Secondly, they do not so much lead us forward as serve to justify the approach by checking that we have not left anything out, that the functions we claim to be structurally stable really are, and that Thom's list is complete. This is an especially crucial point because of the way the theory is applied: much of the power would be lost if we could not be certain that we knew all the possibilities. But if we are prepared to take this assurance on trust – and there appears to be no doubt that the theorems are correct – we lose relatively little by not seeing how it is proved.

Much of the difficulty in establishing the results arises from the fact that we are basing our analysis on Taylor series, even though it is well known that there are smooth functions which are not well approximated by their formal expansions. For example, the function

$$f(x) = \begin{cases} 0, & x = 0, \\ \exp(-1/x^2), & x \neq 0 \end{cases}$$

has a perfectly well-defined expansion, all of the coefficients of which are zero. This is of course the same as the Taylor series for $f(x)=0$, even though the functions coincide at only one point.

It is thus not simply a matter of establishing an isomorphism between functions and Taylor expansions, because no such isomorphism exists. Instead we have to restrict our attention to those functions which are either structurally stable or else structurally unstable but capable of being embedded into structurally stable families with finite numbers of unfolding parameters. It is then possible to show that every such function is, for our purposes, equivalent to a polynomial of the same corank and codimension, and that there is consequently no loss in working in terms of Taylor expansions. (Compare the analysis of the Poston catastrophe machine.) From a practical point of view we lose little by this restriction, as the functions that remain are precisely those which correspond to the structurally stable processes we have set out to study. On the other hand, statements which are true only of subclasses of entities, especially subclasses which are not defined in a simple way, are often much harder to prove than statements which are true of the class as a whole.

3

The seven elementary catastrophes

In this chapter we derive Thom's famous list of seven elementary catastrophes. We have already seen that we need only consider Taylor series in one or two variables; we now have to find all the different cases which can arise with codimension no greater than 4.

We begin with some mathematical preliminaries. First, we have to be clear about what we mean by 'different' cases. The usual statement is that there are seven qualitatively different catastrophes, and when we come to discuss the applications of the theory it will be clear enough what this means. But we need a more precise definition if we are to know what sorts of mathematical operations we may use in the derivations, so we say that two catastrophes are equivalent if one can be transformed to the other by (i) a diffeomorphism of the control variables, and (ii) at each point in the control space a diffeomorphism of the state variables. The resulting family of state-variable diffeomorphisms must be smooth when considered as a function of the control variables.

A *diffeomorphism* is a one-to-one continuous differentiable transformation. It is sometimes useful (though not strictly accurate) to think of two geometric objects as being topologically equivalent, or *homeomorphic*, if one can be continuously deformed into the other without any tearing or pasting together. To the same degree of accuracy we may think of two geometric objects as being diffeomorphic if they are homeomorphic and if, in addition, the deformation involved no creasing or flattening of creases. Thus a sphere, an ellipsoid and a cube are all homeomorphic, but only the first two are diffeomorphic. Note that by insisting on separate diffeomorphisms of state and control variables we ensure that not only the equilibrium surfaces but also the bifurcation sets of two equivalent catastrophes are diffeomorphic.

We also introduce some notation, to prevent the calculations from becoming any messier than they have to be. First, we shall often want to simplify an expression such as

$$(ax + by)(px + qy)^2$$

by letting

$$x' = ax + by, \quad y' = px + qy$$

which gives

$$x'(y')^2$$

or, dropping primes,

$$xy^2.$$

We can express the same thing more concisely by writing

$$(ax + by)(px + qy)^2 \sim xy^2 \text{ using } \Phi: x \mapsto ax + by, y \mapsto px + qy.$$

We use the arrows to define the transformation by giving its effect on x and y. The symbol $\sim$ means 'is equivalent to', which in this chapter will always mean 'is diffeomorphic to', since all our transformations will be diffeomorphisms. Note, by the way, that the given transformation actually carries the simpler form to the more complicated one; it seems nevertheless more natural to write it this way, and in any case there can be no real ambiguity because a diffeomorphism must possess a unique inverse. Where a diffeomorphism is a simple rescaling of one or both variables we shall usually not give it explicitly, and we shall also usually omit giving it a name (in this case Φ) unless we need to.

We shall often write diffeomorphisms in the form

$$\Phi: x \mapsto x + \phi(x, y), \quad y \mapsto y + \psi(x, y),$$

where ϕ and ψ are polynomials. We could, of course, absorb the x and y into the polynomials, but we write Φ in this form to make $\phi = \psi = 0$ correspond to the identity diffeomorphism $x \mapsto x$, $y \mapsto y$. As with the functions themselves, it can be shown that for our purposes there is no loss of generality in defining our diffeomorphisms in terms of polynomials.

Finally, in order to avoid having to keep stating that higher order terms are being neglected, we define the *k-jet* of a function f, which we write as $j^k(f)$, to be the formal Taylor series of f truncated after the kth order term. Thus, for example,

$$j^3(\sin x) = x - x^3/3!.$$

A function is said to be *k-determined* if any function with the same k-jet is of the same type; a function is said to be *finitely determined* if it is k-determined for some finite k. Two functions which are not finitely determined are $f(x)=0$ and the $\exp(-1/x^2)$ we mentioned in the previous chapter.

The singularities

If there is only one essential variable, then we can see at once that there are exactly four singularities with codimension less than or equal to 4:

Singularity	Codimension	Name
x^3	1	fold
x^4	2	cusp
x^5	3	swallowtail
x^6	4	butterfly

The origin of the names will become obvious in the next chapter, when we study the geometry of these catastrophes. The cusp is sometimes called the 'Riemann–Hugoniot' catastrophe; this is the name that Thom originally gave it to acknowledge that it first appeared in the literature in the study of the shock wave that appears in front of an accelerating piston. Catastrophes of corank 1 are referred to collectively as *cuspoids*.

If the corank is 2, then we are dealing with Taylor series in two variables in which the quadratic terms vanish identically, and which therefore begin with a homogeneous cubic. In our new terminology we say that we are considering singularities η such that $j^2(\eta)=0$ and

$$j^3(\eta)=(a_1x+b_1y)(a_2x+b_2y)(a_3x+b_3y),$$

using the fact that a homogeneous polynomial can always be written as the product of linear factors, if we allow the possibility that some of the coefficients may be complex. What we now have to do is to consider the different cases that can arise and derive canonical forms for them. This is exactly analogous to the way in which we take the general quadratic equation $ax^2+2hxy+by^2+2gx+2fy+c=0$ and show that it represents one of a relatively short list of conic sections, each of which can be written in a simple form which is easy to work with. The difference is that in analysing the conics we are allowed only translations and rotations of the coordinate axes because we have to preserve the shapes of the figures; in the present case we have at our disposal the much larger class of diffeomorphisms.

Suppose first that all the coefficients a_i, b_i are real, and that no two of the ratios a_i/b_i are equal. Then

$$j^3(\eta)\sim(ax+by)xy \quad \text{using } x\mapsto a_2x+b_2y,\ y\mapsto a_3x+b_3y$$

$$\sim (x+y)xy$$
$$\sim x(x^2 - y^2) \qquad \text{using } x \mapsto x+y, \; y \mapsto x-y.$$

Thus the canonical form of the *elliptic umbilic* is

$$x^3 - xy^2.$$

Of course $x^2 y + xy^2$ would also do as a canonical form, but it turns out that it is less convenient to use.

Note that out of all possible diffeomorphisms we used only linear transformations. This is because any zero order terms would have violated the condition that $j^2(\eta)$ must vanish, while the effect of higher order terms would have disappeared in the truncation to the 3-jet.

The above reduction to canonical form will fail if not all the a_i, b_i are real, because we are not permitted to use complex coordinate transformations. In this case, since the complex roots of a polynomial equation with real coefficients occur in conjugate pairs, we have

$$j^3(\eta) = (a_1 x + b_1 y)(a_2 x + b_2 y)(\bar{a}_2 x + \bar{b}_2 y)$$

with a_1, b_1 real. It is easily verified that

$$(a_2 x + b_2 y)(\bar{a}_2 x + \bar{b}_2 y)$$
$$= (\mathrm{Re}(a_2)x + \mathrm{Re}(b_2)y)^2 + (\mathrm{Im}(a_2)x + \mathrm{Im}(b_2)y)^2$$

from which it follows that

$$j^3(\eta) \sim (ax + by)(x^2 + y^2)$$
$$\sim x(x^2 + y^2) \qquad \text{using } x \mapsto ax + by, \; y \mapsto bx - ay$$
$$\sim x^3 + xy^2.$$

This form is sometimes used as the canonical one, and it does have the advantage that it is parallel to the canonical form of the elliptic umbilic. We can, however, obtain a somewhat neater form by noticing that

$$(x+y)^3 + (x-y)^3 = 2x^3 + 6xy^2 \sim x^3 + xy^2$$

so that using the transformation $x \mapsto x+y$, $y \mapsto x-y$ we obtain the canonical form of the *hyperbolic umbilic*:

$$x^3 + y^3.$$

So far we have supposed that all three of the ratios a_i/b_i were different. We now consider what happens if $a_1/b_1 = a_2/b_2$ so that

$$j^3(\eta) = (a_1 x + b_1 y)^2 (a_3 x + b_3 y)$$
$$\sim x^2 y$$

which is not finitely determined. We shall be able to justify this claim shortly, but intuitively it seems obvious that one term is not enough to determine the behaviour of a function in two directions, especially since it is only a part of a canonical form that we have already found. There is thus a degeneracy in the 3-jet, and so exactly as in the single variable case we have to go on to the 4-jet, which can be written

$$j^4(\eta) = x^2 y + h(x, y),$$

where $h(x, y)$ is a homogeneous polynomial of degree 4.

We want to get this into a canonical form, but we have to be careful not to do anything that will alter the 3-jet. So we try a diffeomorphism

$$\Phi: \ x \mapsto x + \phi(x, y), \quad y \mapsto y + \psi(x, y),$$

where ϕ and ψ are polynomials. The effect of this on $j^4(\eta)$ is given by

$$\Phi: \ x^2 y + h(x, y)$$
$$\mapsto x^2 y + x^2 \psi + 2xy\phi + 2x\phi\psi + y\phi^2 + \phi^2 \psi + h(x + \phi, y + \psi).$$

If the 3-jet is to be unchanged, then neither ϕ nor ψ can have any zero or first order terms. But if the lowest order terms in ϕ and ψ are quadratic, then

$$h(x + \phi, y + \psi) = h(x, y) + \text{5th and higher order terms.}$$

Hence when we truncate after the 4th order terms we find

$$\Phi: \ x^2 y + h(x, y) \mapsto x^2 y + x^2 \psi + 2xy\phi + h(x, y).$$

We now choose ϕ and ψ to eliminate as much of $h(x, y)$ as possible. We may write

$$h(x, y) = ay^4 + by^3 x + cy^2 x^2 + dyx^3 + ex^4$$

and then by choosing

$$\psi(x, y) = -(cy^2 + dxy + ex^2),$$
$$\phi(x, y) = -\tfrac{1}{2}by^2$$

we show that

$$j^4(\eta) \sim x^2 y + y^4,$$

which is the canonical form of the *parabolic umbilic*.

The elliptic and hyperbolic umbilics both have codimension 3, which is the minimum for a singularity of corank 2. In the case of the parabolic umbilic, the extra condition $a_1 b_2 = a_2 b_1$ raises the codimension to 4. Any further degeneracy in the 3-jet, or the requirement that in the quartic $h(x, y)$ the coefficient of y^4 be zero (which we assumed above it was not) would increase the codimension to 5 or above. The list of singularities of codimension less than or equal to 4 is therefore complete.

Before we go on to derive the unfoldings, there is one further point to mention. The singularity $-x^4$ has precisely the same geometry associated with it as does x^4, and for this reason Thom, who is primarily interested in the form of the bifurcation set, does not distinguish between them. On the other hand, the two do differ in that maxima and minima are interchanged, and this obviously does matter in some applications. For this reason it is sometimes convenient to use a finer classification than Thom's, and to list $-x^4$ separately as the *dual cusp* catastrophe.

Many catastrophes are self-dual. For example, replacing x by $-x$ carries x^3 to $-x^3$, and of course these do have the same form at the origin, viz. a point of inflexion. Of the seven catastrophes in Thom's list, only the cusp, the butterfly and the parabolic umbilic are not self-dual.

The universal unfoldings

Let $\eta(x, y)$ be a finitely determined singularity and let $g(x, y)$ be a polynomial which is close to η. Then g may be written

$$g(x, y) = \eta(x, y) + \sum_i \sum_j \varepsilon_{ij} x^i y^j,$$

where the coefficients ε_{ij} are small and the summations are over all i, j. Now g is clearly a versal unfolding of η, since it includes all types possible among two-variable potentials, but it is not much use as it stands, because it contains far too many terms. The problem is to find the universal unfolding, i.e. the versal unfolding with the smallest number of parameters. We remark that one advantage we have in solving this problem is that at least we will recognize the universal unfolding when we find it, because the number of unfolding parameters will be equal to the codimension of η.

The first step in the process of finding a universal unfolding is the following. Consider the infinitesimal diffeomorphism

$$\Phi: x \mapsto x + \phi(x, y), \ y \mapsto y + \psi(x, y),$$

where ϕ and ψ are polynomials all of whose coefficients are small. The effect of Φ on η is given by

$$\Phi: \eta(x, y) \mapsto \tilde{g}(x, y) = \eta(x, y) + \phi(x, y)\frac{\partial \eta}{\partial x} + \psi(x, y)\frac{\partial \eta}{\partial y}.$$

Because Φ is a diffeomorphism, η and $\tilde{g}$ are of the same type. Hence in the versal unfolding g, the only terms that can produce changes in the type are those that we cannot arrange to occur in $\tilde{g}$ by suitable choice of ϕ and ψ. It follows that the universal unfolding will contain no terms that are multiples either of $\partial \eta/\partial x$ or of $\partial \eta/\partial y$. It will also have no constant term, since such a term, which can be eliminated by a change of origin of g, cannot affect the type.

This is not usually enough to give us the universal unfolding immediately, but it does reduce the problem to manageable proportions. We may look on what we are doing as trying to account for all the terms in g. We eliminate as many as possible by the infinitesimal diffeomorphism, and then use unfolding parameters to deal with any that remain.

As before, we can dispose of the cuspoids quickly. The singularity is in each case of the form x^n. We can find a diffeomorphism to eliminate all terms which are multiples of $\partial\eta/\partial x$, i.e. of x^{n-1}, and so the universal unfoldings are

$$x^3 + ux \qquad\qquad \text{fold,}$$
$$x^4 + ux^2 + vx \qquad\qquad \text{cusp,}$$
$$x^5 + ux^3 + vx^2 + wx \qquad \text{swallowtail,}$$
$$x^6 + tx^4 + ux^3 + vx^2 + wx \quad \text{butterfly.}$$

We now turn to the umbilics, and this time we begin with the hyperbolic umbilic, because it is the easiest. The canonical form of the singularity is $x^3 + y^3$ and so

$$\partial\eta/\partial x \sim x^2, \quad \partial\eta/\partial y \sim y^2.$$

We may therefore omit from the unfolding all monomials which are multiples of either x^2 or y^2. There are only three monomials left, x, y and xy, and since the codimension of the singularity is 3 we know that the universal unfolding is

$$x^3 + y^3 + wxy + ux + vy.$$

Finding universal unfoldings is seldom so straightforward as this, and the best way of keeping track of what is going on is to use 'Siersma's trick'. This consists of writing all possible monomials in x and y in the form of a triangular array:

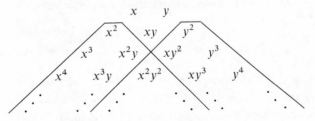

The point is that any monomial in the array is a divisor of any monomial which lies on or below the diagonals drawn down from it. Hence once we have shown that a particular monomial is accounted for we can draw a 'shadow' from it, and then ignore all monomials which lie in the shadow. In the above example we have illustrated the procedure by drawing the shadows from x^2 and y^2, and the three unfolding terms are immediately apparent. The fact that the third row of the array is completely shaded tells us that the singularity is 3-determined.

The elliptic umbilic is somewhat harder. The canonical form of the singularity is $x^3 - xy^2$, which gives

$$\partial\eta/\partial x \sim x^2 - y^2, \quad \partial\eta/\partial y \sim xy,$$

and while we can draw a shadow from xy, it is not obvious what we are to do with $x^2 - y^2$. We can, however, make some progress by noticing that

$$x^3 = x(x^2 - y^2) + y(xy),$$
$$y^3 = -y(x^2 - y^2) + x(xy)$$

which means that we can eliminate any x^3 or y^3 terms. We then draw shadows from x^3 and y^3, and we find that the whole of the third row is now covered, so this singularity too is 3-determined. On the other hand, there are still four monomials not in the shade, whereas we know that the codimension of the singularity is only 3.

Suppose, however, that we have an unfolding term proportional to $x^2 + y^2$. Then since

$$(x^2 + y^2) + (x^2 - y^2) = 2x^2,$$
$$(x^2 + y^2) - (x^2 - y^2) = 2y^2$$

we can use this unfolding term, together with the $x^2 - y^2$ we already have, to eliminate both x^2 and y^2 terms separately. Hence the universal unfolding of the elliptic umbilic may be written as

$$x^3 - xy^2 + w(x^2 + y^2) + ux + vy.$$

We could also have chosen either wx^2 or wy^2 as the quadratic term, but the symmetric form generally turns out to be more convenient.

Note that it is legitimate to combine an unfolding term with a term we obtain from a partial derivative of η since we are free to choose the diffeomorphism Φ to suit our needs. We may not, however, combine two unfolding terms in the same way, as this would imply that the corresponding control variables were not independent.

Finally we come to the parabolic umbilic. The canonical form of the singularity is $y^4 + x^2 y$ so that

$$\partial\eta/\partial x \sim xy, \quad \partial\eta/\partial y \sim x^2 + y^3.$$

Again only xy has an obvious place in the diagram, so we draw a shadow from it and then start looking for combinations that give us more monomials. One such is

$$y^2(xy) - x(x^2 + y^3) \sim x^3$$

so we can draw a shadow from x^3. We now need a shadow down the right hand diagonal, and we would like it to be from y^3 because that would leave four unshaded monomials, and the codimension of the singularity is 4. But the best that we can do directly is to notice that

$$y(x^2 + y^3) - x(xy) = y^4$$

which allows us to draw the shadow only from y^4. Actually this is what we should have expected, because had we been able to cover the third row completely using only monomials based on $\partial \eta / \partial x$ and $\partial \eta / \partial y$ then we would have shown that the singularity was 3-determined, which we know it is not. On the other hand, if we have an unfolding term in x^2, then we can use it not only to eliminate any x^2 term but also, in combination with $y^3 + x^2$, to eliminate any y^3 term as well. Hence we do not need a y^3 term in the unfolding, and the universal unfolding of the parabolic umbilic is

$$y^4 + x^2 y + wx^2 + ty^2 + ux + vy.$$

This completes the derivation of the seven elementary catastrophes and their unfoldings. We recall, however, that during the derivation of the parabolic umbilic we made the claim that the singularity $x^2 y$ is not finitely determined. We can now justify this statement by remarking that the two partial derivatives are equivalent to xy and x^2, and that when we put these into the diagram we leave the right hand diagonal completely uncovered. There is therefore no row which is completely covered, and $x^2 y$ is consequently not finitely determined. It is of infinite codimension.

Higher order catastrophes

Most of the applications of catastrophe theory are based on the seven elementary catastrophes whose canonical forms we have just derived. Since, however, the 5-parameter catastrophes are not really any harder, we sketch the derivations here; the reader may take the detailed calculation of the canonical forms as an exercise to check his understanding of this chapter.

The first of these catastrophes is of corank 1, and is obviously

$$x^7 + sx^5 + tx^4 + ux^3 + vx^2 + wx.$$

This is known as the *wigwam* on account of a shape that appears when the geometry is studied.

If the corank is 2, then the 3-jet can, as we have seen, be written

$$j^3(\eta) = (a_1 x + b_1 y)(a_2 x + b_2 y)(a_3 x + b_3 y).$$

So far we have considered the cases when none of the ratios a_i/b_i were equal, which led to two catastrophes of codimension 3, and when one pair were equal, which led to a catastrophe of codimension 4. If all three of the ratios are equal, then we have two conditions to add to the three we get by requiring that $j^2(\eta)$ vanish, and so we have a catastrophe of codimension 5. The 3-jet is obviously equivalent to x^3; hence we consider singularities η whose 4-jets can be written

$$j^4(\eta) = x^3 + h(x, y)$$

where, as in the case of the parabolic umbilic, $h(x, y)$ is a homogeneous quartic in x and y. This time, however, when we consider the diffeomorphism

$$\Phi: x \mapsto x + \phi(x, y), \quad y \mapsto y + \psi(x, y)$$

we find that while ϕ must still have no zero or first order terms, a linear term in ψ will not affect the 3-jet. This enables us to write the canonical form of the singularity as $x^3 + y^4$, and Siersma's trick then immediately gives us the universal unfolding of the *symbolic umbilic*:

$$x^3 + y^4 + sxy^2 + ty^2 + uxy + vy + wx.$$

The name, incidentally, was chosen by phonetic analogy, and not for any deeper reason.

Any further degeneracy in the 3-jet would increase the codimension, so there are no more cases to consider there. On the other hand, it may turn out that when we are analysing a singularity with a single degeneracy in the 3-jet we find that we do not obtain the parabolic umbilic because the coefficient of the y^4 term is zero. This is an additional condition (though one which might not be obvious in the first place because what has to vanish is not necessarily the y^4 term in the original potential but rather what becomes the y^4 term after some coordinate transformations) and so the codimension of the singularity will be 5. In this case we have to look at the 5-jet, and since the only 5th degree monomial which is not divisible either by xy or by x^2 is y^5, the remaining singularities are $y^5 + x^2y$ and $-y^5 + x^2y$. We can readily verify that these are separate catastrophes, not duals, and then, by using arguments analogous to those we used to derive the unfolding of the parabolic umbilic, we can derive the universal unfoldings of the *second elliptic umbilic* and the *second hyperbolic umbilic*:

$$x^2y \mp y^5 + sy^3 + ty^2 + ux^2 + vy + wx.$$

When the codimension exceeds 5, we run into problems. We find more cuspoids, all of the form x^n, and the so-called conic umbilics, $x^2y \pm y^n$. There are also many others; for one thing, codimension 6 allows catastrophes of corank 3. The real problem, however, is that the method we have been using to group our catastrophes into a small number of classes no longer works.

It is easiest to see what goes wrong by considering the singularities whose 3-jets vanish identically, and whose 4-jets can therefore be written as

$$j^4(\eta) = (a_1 x + b_1 y)(a_2 x + b_2 y)(a_3 x + b_3 y)(a_4 x + b_4 y).$$

Clearly there are a number of possibilities, depending upon how many of the coefficients are complex and what sorts of degeneracies there are. Suppose, however, that all the coefficients are real and that all the ratios a_i/b_i are different. Then if we try to proceed as for the elliptic umbilic we find

$$j^4(\eta) \sim xy(ax + by)(px + qy) \quad \text{using } x \mapsto a_1 x + b_1 y, \; y \mapsto a_2 x + b_2 y$$
$$\sim xy(x + y)(sx + ty)$$
$$\sim xy(x + y)(x + \lambda y)$$

and we are left with a 1-parameter family of catastrophes, rather than the single canonical form we were looking for. The members of the family are not equivalent in the sense that they cannot be transformed into each other by diffeomorphisms. But they are equivalent in the sense that they are all quartics with four distinct real roots. It follows that while the algebraic codimension of this singularity is 8, the topological codimension is only 7.

The non-degenerate quartics in two variables form a set of functions of codimension 7. This set can be split into three subsets, with four, two and no real roots, respectively, and each of these subsets can be partitioned into families. Every member of one of these families is diffeomorphic to every other member of the same family, and each family is of codimension 8. Canonical forms of the three types of the *double cusp* catastrophe can be shown to be

$$x^4 + y^4 \qquad x^4 - y^4 \qquad x^4 - 6x^2 y^2 + y^4$$

though, as is clear from the foregoing, these are not derived in quite the same way as the canonical forms for the singularities of codimension not greater than 5. See Poston & Stewart (1976, 1978a) or Zeeman (1976b) for more on the double cusp.

4

The geometry of the seven elementary catastrophes

Now that we have the list of seven elementary catastrophes, we have to discover their properties. This is a comparatively straightforward task, and we shall carry out almost all the necessary calculations explicitly.

What we have to do is precisely what we did when we analysed the catastrophe machines in the first chapter. Given a potential, V, we define the *equilibrium surface*, M, by the equation

$$\nabla_x V = 0,$$

where the subscript x indicates that the gradient is with respect to the state variables only. This surface is made up of all the critical points of V, i.e. all the equilibria (stable or otherwise) of the system. We denote it by M to indicate that it is a manifold, a well-behaved smooth surface. It is not, by the way, obvious that M must be a manifold, but it can be proved that it is.

Next we find the *singularity set*, S, which is the subset of M which consists of all the degenerate critical points of V. These are the points at which $\nabla_x V = 0$ and also

$$\Delta \equiv \det\{H(V)\} = 0,$$

where $H(V)$ is the Hessian of V, the matrix of second order partial derivatives which we defined in Chapter 2. We then project S down into the control space C (by eliminating the state variables from the equations which define it) to obtain the *bifurcation set*, B, which is the set of all points in C at which changes in the form of V occur. Finally we determine the form of V at every point in C; this is easier than it sounds because since changes can occur only on B it is sufficient to consider only one point within each of the regions into which B divides C.

The fold
This is naturally the easiest catastrophe to analyse. The potential is

$$V(x) = x^3 + ux$$

so the phase space is two-dimensional. The equilibrium surface M is the curve

$$3x^2 + u = 0. \tag{1}$$

The singularity set S is the subset of M for which the equation

$$6x = 0 \tag{2}$$

is also satisfied, i.e. the single point $(0, 0)$. The bifurcation set B is the projection of this onto the control space (i.e. onto the line $x = 0$) and is therefore the single point $u = 0$. See Fig. 4.1.

The bifurcation set divides the control space into two regions, the positive and negative u-axes. If $u > 0$, then equation (1) has no real solutions and V has no critical points. If $u < 0$, on the other hand, V has two critical points, a minimum and a maximum, and so there are two equilibria, one stable and one unstable. On B (i.e. when $u = 0$) these merge into a point of inflexion. A system for which this V is the potential cannot persist stably with $u > 0$, and we express this by saying that on the positive u-axis the only possible regime is the empty regime.

The cusp (or Riemann–Hugoniot)

We have analysed this before, but for the sake of completeness we include it here, this time using the standard format. The potential is

$$V(x) = x^4 + ux^2 + vx$$

Fig. 4.1. The equilibrium surface and bifurcation set of the fold catastrophe.

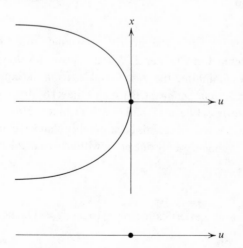

so the phase space is three-dimensional. The equilibrium surface M is given by

$$4x^3 + 2ux + v = 0 \tag{1}$$

and the singularity set is the subset of M for which the equation

$$12x^2 + 2u = 0 \tag{2}$$

is also satisfied. We find the bifurcation set by eliminating x from (1) and (2), obtaining

$$8u^3 + 27v^2 = 0.$$

We shall not repeat the determination of the forms of V, but we recall that within the cusp there are two minima separated by a maximum, whereas outside it there is a single minimum (Fig. 4.2).

For the cusp, as indeed for all the elementary catastrophes, there is no generally accepted notation. Different authors use different symbols for the unfolding parameters, and some prefer to write V as $\frac{1}{4}x^4 + \frac{1}{2}ux^2 + vx$ so as not to have numerical factors in the equation of M. On the other hand, most writers do seem to follow Zeeman (1976a) in referring to the

Fig. 4.2. The equilibrium surface and bifurcation set of the cusp catastrophe.

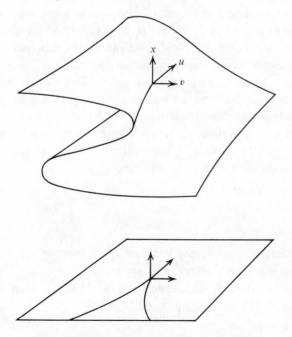

coefficients of x and x^2 as the *normal* and *splitting* factors, respectively. The names reflect the fact that when $u > 0$ then changes in v produce only smooth changes in x – which we may call 'normal' behaviour – but when we decrease u to negative values we 'split' M and discontinuities in x can occur.

The swallowtail

The potential is

$$V(x) = x^5 + ux^3 + vx^2 + wx,$$

so the phase space is four-dimensional, which means that we cannot draw a diagram equivalent to Fig. 4.2. The equilibrium surface M is the hypersurface

$$5x^4 + 3ux^2 + 2vx + w = 0 \tag{1}$$

and the singularity set is the subset of M for which the equation

$$20x^3 + 6ux + 2v = 0 \tag{2}$$

also holds.

It is possible to eliminate x directly from (1) and (2) and hence find the equation of the bifurcation set B, which is a surface in the three-dimensional control space C. Since we are concerned only with the qualitative behaviour of the system, and therefore want primarily to be able to sketch B, a different approach turns out to be more effective. Let C_u be a plane $u = $ constant in C, and let B_u be the intersection of C_u with B. Then B_u will be a curve in C, and if we can sketch this curve for all values of u we can build up the complete surface B.

Even with u held constant it is better not to eliminate x from the equations but to consider it as a parameter along B_u. We then remark that (2) implies that v is an odd function of x and that this, together with (1), implies that w is an even function of x. Hence w is an even function of v, and so for any u the curve B_u is symmetric about the w-axis.

Next we differentiate (1) and (2), obtaining

$$dw/dx = -2x \, dv/dx \tag{3}$$

and

$$dv/dx = -(30x^2 + 3u), \tag{4}$$

the rest of the terms in (3) having vanished on account of (2). We now have to consider the cases $u > 0$ and $u < 0$ separately.

If u is positive, then dv/dx cannot vanish. Hence v is a strictly monotone function of x and the equation

$$dw/dv = -2x \tag{5}$$

is valid everywhere. Moreover, equation (2) implies that $xv < 0$, with equality only when $x = v = 0$, at which point w also vanishes. It follows that B_u is smooth, that w is large when $|x|$ is large, and that the sign of dw/dv is the same as that of v, vanishing only at the origin. This enables us to draw Fig. 4.3a.

If u is negative, then the situation is more complicated. We see from (4) that dv/dx vanishes for two real values of x, $\pm\sqrt{(-0.1u)}$. Consequently dw/dx vanishes for three values of x, these two together with $x = 0$ as before, and it follows that B_u has a critical point at $x = 0$ and cusps at the other two points.

To determine the type of the critical point, we notice that equation (2) implies that for $|x| < \pm\sqrt{(-0.3u)}$ the product xv cannot be negative. Since x and v also vanish together, it follows that if v is small and positive so is x, and dw/dv is then negative. This, together with the fact that B_u is symmetric about the w-axis, establishes that the critical point is a relative maximum.

Finally, we note that if $v = 0$ then either $x = 0$ or $x = \pm\sqrt{(-0.3u)}$. We have just seen that $x = 0$ corresponds to a maximum at the origin, and on substituting into equation (1) we find that both the other roots give $w = 9u^2/20$. Hence B_u has a point of self-intersection on the positive w-axis. We then check that $|x|$ large implies that both $|v|$ and w are also large and then, using the values of the parameter x to tell us that the order of the points we have found is: self-intersection, cusp, maximum, cusp, self-intersection, we can draw Fig. 4.3b. And since the equation of the line of points of self-intersection is the parabola

$$w = 9u^2/20, \quad v = 0,$$

Fig. 4.3. Cross sections of the bifurcation set of the swallowtail: (a) $u > 0$, (b) $u < 0$.

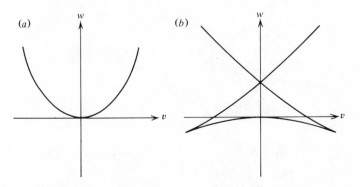

we can put the curves B_u together to form the surface B shown in Fig. 4.4. The origin of the name 'swallowtail' is now apparent.

To find the form of the potential in each of the three regions into which B divides C, it is sufficient to consider points for which $v = 0$ and $u < 0$. Then the solution of equation (1) is

$$x^2 = \tfrac{1}{10}(-3u \pm \sqrt{(9u^2 - 20w)}).$$

There are three cases:

(a) $w > 9u^2/20$ Equation (1) has no real roots and V has no critical points.

(b) $0 < w < 9u^2/20$ Because $\sqrt{(9u^2 - 20w)}$ is real and less than the real and positive $-3u$, both solutions for x^2 are real and positive and V has four critical points, two maxima and two minima.

(c) $w < 0$ Both solutions for x^2 are real but one is negative. Consequently V has only two critical points, one maximum and one minimum.

Fig. 4.4. The bifurcation set of the swallowtail. After Bröcker & Lander (1975).

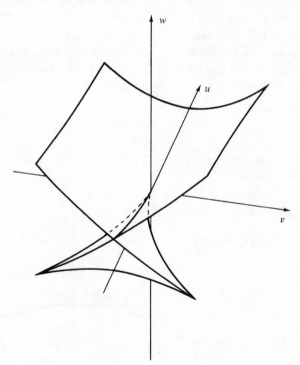

Thus in Fig. 4.4 there is no stable equilibrium possible above the surface, one stable equilibrium below the surface, and two within the swallowtail.

The elliptic umbilic

The potential is

$$V(x, y) = \tfrac{1}{3}x^3 - xy^2 + w(x^2 + y^2) - ux + vy.$$

We have rescaled the variables so as to introduce a factor $\tfrac{1}{3}$ into the first term because it makes the calculations much neater; the qualitative results are of course not affected by this. The phase space is now five-dimensional, but the control space is still only three-dimensional, so we shall still be able to draw it. The equilibrium surface M is the three-dimensional hypersurface whose equations are

$$x^2 - y^2 + 2wx - u = 0, \tag{1a}$$

$$-2xy + 2wy + v = 0, \tag{1b}$$

and the singularity set S is the subset of M for which also

$$\begin{vmatrix} 2x + 2w & -2y \\ -2y & -2x + 2w \end{vmatrix} = 0,$$

i.e.

$$\Delta = 4(w^2 - x^2 - y^2) = 0. \tag{2}$$

We now proceed in much the same way as before. Instead of finding the equation of B directly, by eliminating x and y from the equations of S, we consider planes $w = $ constant and try to sketch the curves B_w. From equation (2) we see that if w is constant we may write

$$x = w \cos \theta. \quad y = w \sin \theta$$

which when substituted into equations (1) yields equations for u and v in terms of a single parameter, θ:

$$u = w^2(\cos 2\theta + 2 \cos \theta),$$

$$v = w^2(\sin 2\theta - 2 \sin \theta).$$

If $w = 0$, then B_w is the single point $u = v = 0$. If $w \neq 0$, then we find

$$du/d\theta = 0 \quad \text{when} \quad \theta = 0, \ \pi, \pm 2\pi/3$$

and

$$dv/d\theta = 0 \quad \text{when} \quad \theta = 0, \ \pm 2\pi/3.$$

Hence there are cusps at

$$(3w^2, 0), \left(-\frac{3}{2} w^2, \frac{3\sqrt{3}}{2} w^2\right), \left(-\frac{3}{2} w^2, \frac{-3\sqrt{3}}{2} w^2\right)$$

and a vertical tangent at $(-w^2, 0)$. It is now easy to draw B_w (Fig. 4.5), and since the lines of cusps are clearly parabolas we can draw the complete bifurcation set B as well (Fig. 4.6).

There are again three regions within which we have to determine the form of V. For the two within the cusped cones we can simplify the calculations by choosing sample points on the w-axis. The equations of M then reduce to

$$x^2 - y^2 + 2wx = 0,$$
$$y(w - x) = 0.$$

There are four solutions:

$$x = w, \; y = \pm w\sqrt{3}; \quad x = y = 0; \quad x = -2w, \; y = 0.$$

All except $x = y = 0$ make the discriminant, Δ, negative, and so are saddle points. If $x = y = 0$ then

$$\Delta = 4w^2 > 0 \quad \text{and} \quad \partial^2 V/\partial x^2 = w.$$

Hence the remaining critical point is a minimum if $w > 0$ and a maximum if $w < 0$. Inside one of the cones, therefore, there are three saddles and a

Fig. 4.5. Cross section of the bifurcation set of the elliptic umbilic. After Bröcker & Lander (1975).

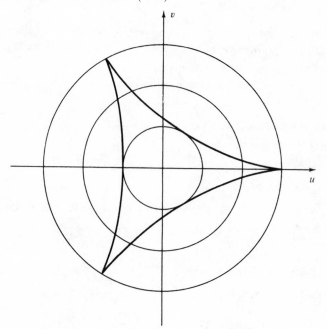

minimum, whereas inside the other there are three saddles and a maximum.

In the remaining region it is convenient to consider the particular point $(u, v, w) = (0, 2, 0)$. At this point the equations of $M'$are

$$x^2 = y^2 \quad \text{and} \quad xy = 1$$

and these have two solutions, $x = y = \pm 1$. In either case, $\Delta = -8$, so at every point in this region the potential has two critical points, both saddles. Thus in only one of the three regions is anything other than the empty regime possible.

The hyperbolic umbilic

The potential is

$$V(x, y) = x^3 + y^3 + wxy - ux - vy,$$

so again we have a five-dimensional phase space and a three-dimensional control space. Note that we have changed the signs of u and v to make the calculations neater. The equations of M are now

$$3x^2 + wy - u = 0, \tag{1a}$$

$$3y^2 + wx - v = 0, \tag{1b}$$

Fig. 4.6. The bifurcation set of the elliptic umbilic. After Bröcker & Lander (1975).

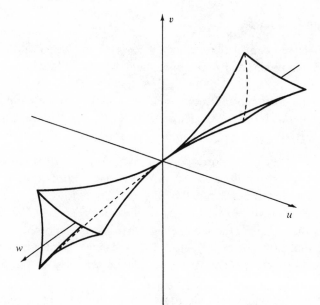

and the singularity set is the subset of M for which also

$$\begin{vmatrix} 6x & w \\ w & 6x \end{vmatrix} = 0,$$

i.e.

$$\Delta = 36xy - w^2 = 0. \tag{2}$$

Again the simplest way forward is to consider sections $w = $ constant. We remark first that if $w = 0$ then either x or y must vanish. If $x = 0$, then (1a) implies $u = 0$ while (1b) requires that v be positive. From this, and the similar result for $y = 0$, we see that B_0 consists of the positive u- and v-axes.

Now suppose $w \neq 0$. We use equation (2) to express y in terms of x and then substitute into (1) to obtain parametric equations for u and v:

$$u = 3x^2 + w^3/36x, \tag{3a}$$

$$v = 3w^4/36^2 x^2 + wx. \tag{3b}$$

If $|x|$ is very small, then both $|u|$ and $|v|$ are very large. On the other hand, v is positive whether x is small and positive or small and negative, whereas u changes sign (since it has the same sign as x if $w > 0$ and the opposite sign if $w < 0$). Consequently B_w is not a continuous curve; it is made up of two disjoint pieces.

We next differentiate equations (3) with respect to x:

$$du/dx = 6x - w^3/36x^2,$$

$$dv/dx = -6w^4/36^2 x^3 + w.$$

Both derivatives vanish if and only if $x = w/6$, so B_w has no relative maxima or minima and only one cusp, which is located at $(\frac{1}{4}w^2, \frac{1}{4}w^2)$.

If $w > 0$, the portion of B_w corresponding to $x < 0$ is smooth and has no stationary points. It crosses the u-axis when $x = -w/(3.4^{1/3})$, i.e. when

$$v = \tfrac{1}{3}w^2(1/16^{4/3} - 1/4^{1/3}) < 0.$$

The portion of the curve corresponding to $x > 0$ has a cusp but neither stationary points nor intersections with the axes. From this, together with the fact that the symmetry of V with respect to x and y implies symmetry of B_w with respect to the line $u = v$, we can draw Fig. 4.7.

If $w < 0$, the picture is the same, except that it is the smooth portion of the curve that now corresponds to $x > 0$. The lines of intersection of B with $u = v$ are again parabolas, so we can immediately sketch B itself (Fig. 4.8).

The bifurcation set divides the control space into four regions, and we can examine three of them by considering points on the line

$$u = v, \quad w = 1.$$

Substituting these equations into equations (1) gives

$$3x^2 + y = 3y^2 + x,$$

i.e.

$$(x - y)(x + y - \tfrac{1}{3}) = 0,$$

so that

$$x = y \quad \text{or} \quad x + y = \tfrac{1}{3}.$$

Fig. 4.7. Cross section of the bifurcation set of the hyperbolic umbilic.

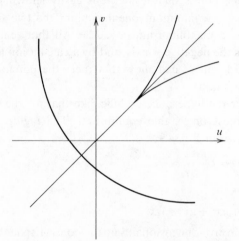

Fig. 4.8. The bifurcation set of the hyperbolic umbilic.

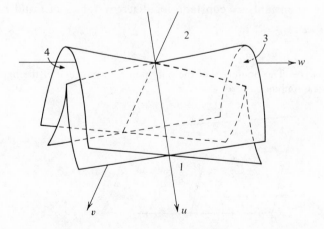

These are satisfied simultaneously only at the cusp points, and we are not considering these as they lie on B itself.

If $x = y$, then equation (1a) gives

$$3x^2 + x - u = 0,$$

which has two real roots if and only if $u > -1/12$. If $x + y = \frac{1}{3}$, then equation (1a) gives

$$3x^2 - x + \tfrac{1}{3} - u = 0,$$

and this has two real roots if and only if $u > \frac{1}{4}$. Since $u = v = -\frac{1}{12}$ is on the smooth portion of B_1 and $u = v = \frac{1}{4}$ is its cusp point, we see that there are four critical points in region 1, two in region 2, and none in region 3. By choosing suitable sample points on the line it is easily shown that in region 1 the potential has one minimum, one maximum and two saddles, while in region 2 it has a maximum and a saddle. All that remains is region 4, which includes the negative w-axis, and by a calculation like the one we have just carried out we can show that there the potential has one saddle and one minimum.

Thus the potential for the hyperbolic umbilic, like that for the elliptic umbilic, can have at most one stable equilibrium. The region within which this exists is shown in Fig. 4.9.

The butterfly

The potential is

$$V(x) = x^6 + tx^4 + ux^3 + vx^2 + wx.$$

The phase space is again five-dimensional, but the control space is now four-dimensional, so we cannot draw the bifurcation set. The most useful approach appears to be to sketch the curves B_{tu}, i.e. the sections of B by the plane $t = $ constant, $u = $ constant, for different values of t and u. (We

Fig. 4.9. The boundary of the only non-empty regime of the hyperbolic umbilic.

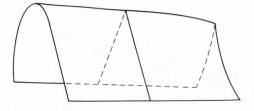

recall that in R^4 a plane is of codimension 2 and is therefore specified by two equations.) In fact we shall perform the calculation explicitly only for $u=0$, as this is sufficient to demonstrate the characteristic features of the butterfly catastrophe.

The equilibrium surface M is the hypersurface

$$6x^5 + 4tx^3 + 3ux^2 + 2vx + w = 0, \tag{1}$$

and the singularity set is the subset of M for which also

$$30x^4 + 12tx^2 + 6ux + 2v = 0. \tag{2}$$

We hold t and u fixed and use x as a parameter along B_{tu}:

$$-v = 15x^4 + 6tx^2 + 3ux, \tag{3}$$
$$w = 24x^5 + 8tx^3 + 3ux^2. \tag{4}$$

If $t=u=0$, we can eliminate x:

$$-(v/15)^5 = (w/24)^4$$

so B_{00} is a simple cusp. To see what happens for other values of t and u we differentiate (3) and (4):

$$-dv/dx = 60x^3 + 12tx + 3u, \tag{5}$$
$$dw/dx = 120x^4 + 24tx^2 + 6ux. \tag{6}$$

We note that dw/dx vanishes alone if $x=0$ and that both derivatives vanish if

$$20x^3 + 4tx + u = 0. \tag{7}$$

Equations (5) and (6) have no other zeros, so B_{tu} can have a vertical tangent (which must be at the origin) or cusps, but no horizontal tangents.

Since equation (7) is a cubic it must always have at least one real root, and B_{tu} must therefore always have at least one cusp. There will be three cusps if all the roots of (7) are real, and the condition for this is

$$u^2 + 4(4t/3)^3 < 0.$$

This condition cannot be satisfied if t is positive, and so t is called the 'butterfly' factor to remind us that it is essentially by changing this variable that we progress from a simple cusped curve to one with the butterfly-like shape. The variable u is called the 'bias' factor because B_{tu} is symmetric about the v-axis only if $u=0$ (by an argument similar to that used to establish the symmetry of the swallowtail). And w and v are called the 'normal' and 'splitting' variables, respectively, because they play the same roles as the normal and splitting variables of the cusp catastrophe.

We now set $u=0$, so that equations (3)–(6) become

$$-v = 15x^4 + 6tx^2, \tag{8}$$

$$w = 24x^5 + 8tx^3, \tag{9}$$

$$-dv/dx = 60x^3 + 12tx, \tag{10}$$

$$dw/dx = 120x^4 + 24tx^2. \tag{11}$$

If $t>0$, then it is easily seen that v is always negative, that there is a cusp at the origin (and hence no separate vertical tangent) and that B_{tu} is a simple cusped curve.

If $t<0$, then the two derivatives vanish together at $x=0$ and $x=\pm\sqrt{(-t/5)}$, so there are now three cusps, but again there is no separate vertical tangent. By direct substitution into (8) and (9) we can show that the two cusps not at the origin are in the upper half plane. Finally, we locate the intersections with the axes:

$v=0$ implies $x=0$ or $x^2 = -6t/15$,

$w=0$ implies $x=0$ or $x^2 = -t/3$.

Fig. 4.10. Cross section of the bifurcation set of the butterfly with $u=0$ and $t<0$. The form of the potential $V(x)$ in each of the regions is shown. The minima have been labelled to assist in predicting what will happen as the control trajectory crosses different parts of the bifurcation set.

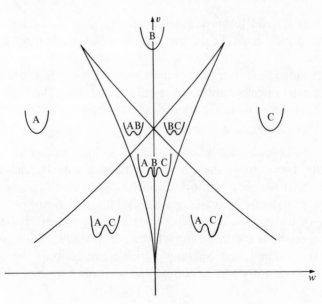

Apart from the origin, which we knew about already, we find that there are two intersections with the w-axis but only one with the v-axis, as the two values $\pm\sqrt{(-t/3)}$ give the same (positive) value of v, viz. $t^2/3$. Hence there is a point of self-intersection on the v-axis. After checking that v and w have the indicated behaviour for large values of $|x|$, we can now draw Fig. 4.10.

Fig. 4.11. Cross sections of the bifurcation set of the butterfly for different values of u and t. After Bröcker & Lander (1975).

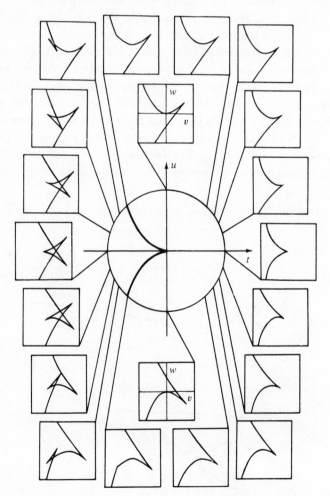

In finding the form of the potential in the different regions we may start by taking $u = w = 0$, so that the equation of M becomes

$$6x^5 + 4tx^3 + 2vx = 0.$$

One root of this is obviously $x = 0$, and the other four are given by

$$x^2 = \tfrac{1}{3}(-t \pm \sqrt{(t^2 - 3v)}).$$

If $t > 0$, then x^2 has a positive real value if and only if $v < 0$. Hence there are three equilibria (two stable, one unstable) within the cusp, but only one (stable) equilibrium outside it, exactly as for the cusp catastrophe.

　　　If $t < 0$, there are three cases:

(a) $v < 0$　　　　three equilibria, two stable and one unstable,

(b) $0 < v < t^2/3$　five equilibria, three stable and two unstable,

(c) $v > t^2/3$　　　one stable equilibrium.

These results are easily established by arguments exactly similar to those we employed when discussing the swallowtail catastrophe. We can deal with the two remaining regions without performing any calculations by remarking that when we cross the bifurcation set of a cuspoid, then in general (by which we mean if we do not cross it at a special point such as a point of self-intersection) we either add or subtract a pair of equilibria, one stable and one unstable. The two regions in question each share ordinary boundaries with a region with five critical points and a region

Fig. 4.12. The equilibrium surface of the butterfly with $u = 0$ and $t < 0$.

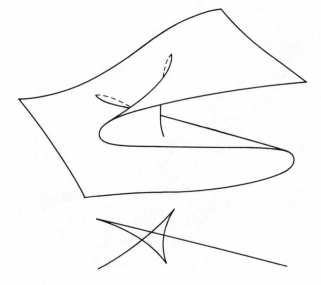

Fig. 4.13. Cross sections of the bifurcation set of the parabolic umbilic for different values of *w* and *t*. After Bröcker & Lander (1975).

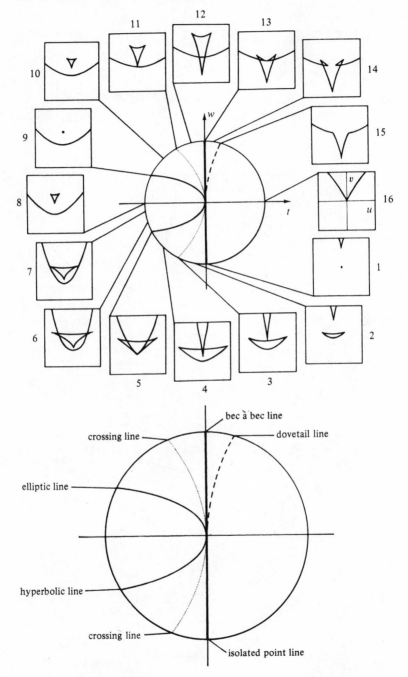

Fig. 4.14. The numbers of local regimes for different values of *w* and *t* (cf. Fig. 4.13). After Bröcker & Lander (1975).

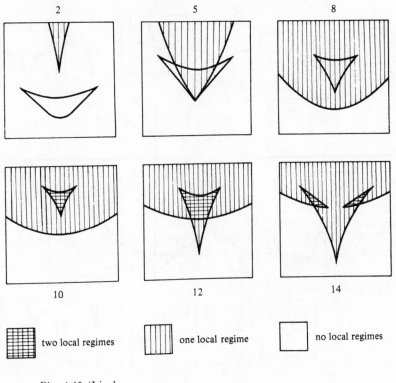

 two local regimes   one local regime   no local regimes

Fig. 4.15. 'Lips'.

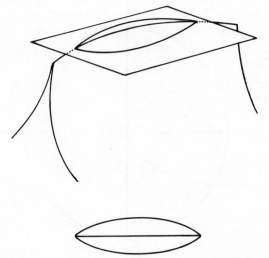

with only one, and it follows that within each of these regions there are three equilibria, two stable and one unstable.

In Fig. 4.11 we show B_{tu} for different values of t and u. We note that the 'pocket' with three stable equilibria appears as we cross a curve in the t–u plane whose equation we know from equation (8) must be

$$u^2 + 4(4t/3)^3 = 0.$$

Hence even if the butterfly factor t is negative, the pocket will not be present if the absolute value of the bias factor u is too large; this observation is important in some applications. Finally, in Fig. 4.12 we show the equilibrium surface M as a function of x, v, w for $u = 0$ and $t < 0$. This illustrates the similarity between the butterfly and the cusp, and of course for $t > 0$ the surfaces are equivalent.

Fig. 4.16. 'Bec-à-bec'.

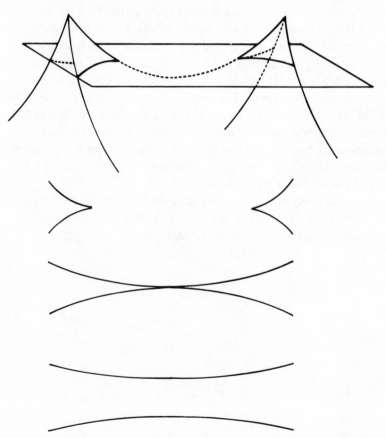

The parabolic umbilic

We take the potential as

$$V(x, y) = y^4 + x^2y + wx^2 + ty^2 - ux - vy.$$

The equilibrium surface M is given by the equations

$$2xy + 2wx - u = 0, \tag{1a}$$

$$x^2 + 4y^3 + 2ty - v = 0, \tag{1b}$$

and the singularity set S is the subset of M for which also

$$\begin{vmatrix} 2y + 2w & 2x \\ 2x & 12y^2 + 2t \end{vmatrix} = 0,$$

i.e.

$$(y + w)(6y^2 + t) = x^2. \tag{2}$$

There does not appear to be a quick way to sketch the bifurcation set, although in principle the parabolic umbilic is no more difficult than the butterfly, so we omit the calculations completely and display without any analysis Figs. 4.13 and 4.14, which are originally due to Chenciner. We recognize in the figure all the catastrophes of codimension less than four: the cusp (16), the swallowtail (14), the elliptic umbilic (10), and the hyperbolic umbilic (5). There are also two unfamiliar configurations, the 'lips' (2) and the 'bec-à-bec' (13), but these are not new catastrophes which we have somehow overlooked. They arise from the fact that in three dimensions we can have not just a cusp, whose bifurcation set is two-dimensional, but a whole line of cusps. If we take sections of this by a plane, then, depending on which way the line curves, we obtain the two additional shapes. See Figs. 4.15 and 4.16.

More on the analysis of the parabolic umbilic can be found in Thom (1972) or Poston & Stewart (1978a). Godwin (1971) has drawn a number of three-dimensional sections of the bifurcation set.

5

Applications in physics

An important feature of catastrophe theory is that it can be used not only in many different problems, but also in many different ways. In an essay entitled 'The two-fold way of catastrophe theory', Thom (1976) has characterized the two ends of the spectrum of the theory's applications as the 'physical' and the 'metaphysical':

> Either, starting from known scientific laws (from Mechanics or Physics) you insert the catastrophe theory formalism (eventually modified) as a result of these laws: this is the physical way. Or, starting from a poorly understood experimental morphology, one postulates 'a priori' the validity of the catastrophe theory formalism, and one tries to reconstruct the underlying dynamic which generates this morphology: this is the 'metaphysical' way. Needless to say, the second way seems to me more promising than the first, if less secure...

It seems natural in a textbook to begin with the more secure examples, and so in this chapter we shall be discussing three applications of catastrophe theory in physics. Because the dynamics are known, almost all the calculations we shall perform will be standard, yet in each case catastrophe theory throws some new light on the problem. In return, these examples contribute to our understanding of catastrophe theory by serving as relatively straightforward illustrations, and also by showing how the range of applicability of the theory extends far beyond systems with gradient dynamics.

Caustics

For the study of many optical phenomena we may ignore the wave nature of light and consider the energy to be transported along curves known as light rays. If we fix our attention on a small set of neighbouring rays, which we call a *pencil* of rays, we find that the intensity of the light is inversely proportional to the cross-sectional area of the pencil. Thus if the rays are parallel the intensity is constant, whereas if the rays are diverging from a point source we obtain the familiar inverse square law of intensity.

It may happen that the rays of a pencil are somehow focussed, so that instead of filling a region of space they are concentrated onto a surface, or even a line or a point. The cross section then has zero area and so, according to geometrical optics, the intensity should be infinite. In fact this is not quite true, as is obvious on physical grounds and as a more accurate analysis confirms, but the intensity can indeed be very great: sufficient to burn a piece of paper, for example. It is for this reason that such surfaces (and also their curves of intersection with the screens with which we observe them) are called *caustics*.

One of the simplest ways of producing a caustic is to allow the light from the sun to fall onto a nearly full cup of coffee. Fig. 5.1*a* illustrates

Fig. 5.1. An easily observed caustic. After Zeeman (1976*a*).

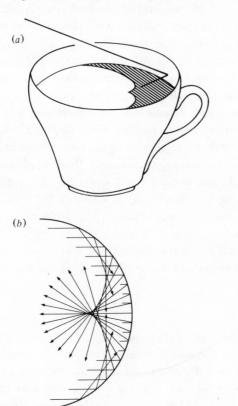

(*a*)

(*b*)

the experimental set-up and what will be observed. This particular caustic is also easy to analyse. Let the inner surface of the cup be represented by the unit circle, and let the incident rays be parallel to the x-axis. If a ray is incident on the cup at the point Q, whose coordinates we may take to be (cos θ, sin θ), then since the angle of reflexion is equal to the angle of incidence we can easily find the equation of the reflected ray from Q:

$$(y - \sin \theta)\cos 2\theta = (x - \cos \theta)\sin 2\theta.$$

Considered as a family of equations with θ as parameter, this represents all the reflected rays.

The caustic is the envelope of this family, i.e. the curve which is tangent to every member of the family. Now the equation of the envelope of a one-parameter family of curves $f(x, y, \theta) = 0$ is found by eliminating the parameter θ from the equations $f = 0$ and $\partial f / \partial \theta = 0$. If this is not easily done, an alternative procedure is to solve the two equations for x and y to obtain the parametric equations of the envelope, which in this case turn out to be

$$x = \cos \theta - \tfrac{1}{2}\cos \theta \cos 2\theta,$$
$$y = \sin \theta - \tfrac{1}{2}\cos \theta \sin 2\theta.$$

These are the equations of a curve known as the nephroid, which is shown in Fig. 5.1. The figure also illustrates how the density of reflected rays is much greater near the envelope than elsewhere, which accounts for the greater intensity. Note that throughout this calculation we have been surpressing the z-coordinate; the equation we have derived is really that of a cylinder with the nephroid as cross section, and what we observe is the intersection of this cylinder with a plane $z = $ constant, i.e. with the surface of the coffee.

Up to this point the analysis is standard, and fuller accounts can be found in (preferably older) texts on optics and on differential equations. To see how catastrophe theory enters into the problem, we repeat the calculation of the equations of the reflected rays, but this time using the somewhat misleadingly named 'principle of least time' due to Fermat.

Consider any point $P(x, y)$ within the unit circle, and let $Q(\cos \theta, \sin \theta)$ be any point on the circle (Fig. 5.2). In general, a ray reflected at Q will not pass through P. The exceptions will be those points Q (i.e. those values of θ) for which the length of the path from the source to P via Q has a stationary – not necessarily minimum – value. If the source is at $x = -d$, and if the incident rays are assumed parallel to the x-axis, the length of this path is

$$g(x, y, \theta) = d + \cos\theta + [(x - \cos\theta)^2 + (y - \sin\theta)^2]^{1/2}.$$

The condition for a stationary value is of course $\partial g/\partial\theta = 0$, and so the family of reflected rays is given by

$$x \sin\theta - y \cos\theta = \sin\theta[(x - \cos\theta)^2 + (y - \sin\theta)^2]^{1/2},$$

which, after some straightforward calculation, can be shown to be the same family $f(x, y, \theta) = 0$ as we obtained by the other method, as indeed it should be. It follows that the equation of the caustic is to be found by eliminating the parameter θ from the equations

$$\partial g/\partial\theta = 0, \quad \partial^2 g/\partial\theta^2 = 0.$$

But these are precisely the equations which define the bifurcation set of g, taking θ as the state variable and x and y as the control variables. Hence the caustic is the bifurcation set for the system with the path length playing the role of the potential. Since there are only two control variables we expect that the catastrophe is a cusp, and by expanding g as a Taylor series in θ about $(\frac{1}{2}, 0, 0)$ we can show that this is in fact correct.

The line of argument we have used does not depend on the shape of the surface off which the light is being reflected, and would apply equally in cases in which the focussing was by refraction. On the other hand, we were able to treat the problem as one of simple minimization only because it was possible to specify a path uniquely by a single parameter. In general, optical paths are determined by finding stationary values of the integral $\int n \, ds$, where n is the index of refraction of the medium. It can,

Fig. 5.2.

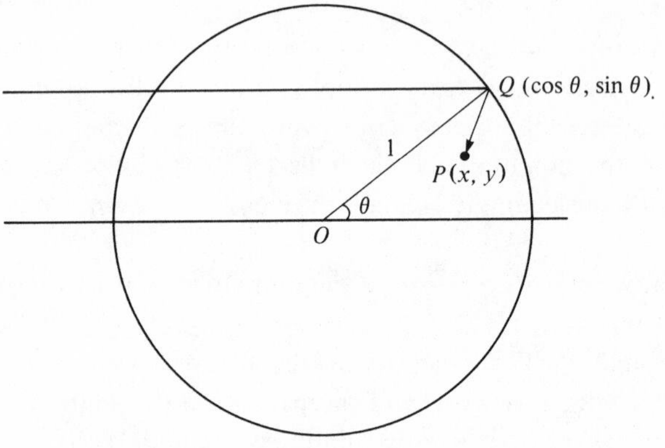

however, be shown that the caustics are still bifurcation sets. Since caustics are surfaces in R^3, there can be no more than three control variables, and so the only structurally stable caustics we observe are folds, cusps, and sections of the swallowtail and the elliptic and hyperbolic umbilics (Berry, 1976).

This result makes it possible to list all the caustics which will be observed in nature. It also helps us with some problems in the understanding of catastrophe theory.

First of all, the system is not derivable from a potential, in the usual sense of the word. It is instead one of a much larger class of systems governed by a variational principle. The significance of a stationary path is not that it expresses some fundamental optimization principle, but rather that it ensures that light from neighbouring paths does not interfere destructively, as it normally does. It is for this reason that maximum-length and minimum-length paths are counted equally. Secondly, the analysis is valid only up to a certain point; for example, we do not actually observe infinite intensities. But this is a consequence of our decision to analyse the system using geometrical optics, and simply illustrates how whether we use catastrophe theory or not we have to be clear about the level of detail at which we intend to work. Finally, while we now know the complete set of caustics which occur in nature, we may find others in artificial systems, such as optical equipment. The whole point of high-precision workmanship, after all, is to construct systems which are not structurally stable.

Non-linear oscillations

A differential equation which is familiar to almost anyone who has studied mathematics is that of the harmonic oscillator,

$$\ddot{x} + x = 0,$$

where, as is customary, a dot indicates differentiation with respect to time. This equation is used to model systems such as a spring or a pendulum performing small oscillations. If we choose the origin of t such that $x(0) = 0$, then the solution is $x = A \cos t$, where A is a constant. (The time variable has been scaled so as to make the angular frequency equal to unity.)

The above equation is, however, structurally unstable, since almost any equation 'near' it will not have strictly periodic solutions. It is also physically unrealistic, because it corresponds to a perfectly conservative system with no energy losses. We can improve it by adding a term $-k\dot{x}$ to the right hand side to provide damping (for a mechanical system this represents a frictional force proportional to the velocity), but then if we want the oscillations to persist we have to supply an energy source in the form of a 'driver', which we shall take to be a periodic force. This leads us to the general linear oscillator

$$\ddot{x} + k\dot{x} + x = F \cos \Omega t, \qquad k > 0.$$

The solution of this equation is

$$x = e^{-kt/2}(A \sin \alpha t + B \cos \alpha t) + \frac{F\cos(\Omega t - \phi)}{\sqrt{((1 - \Omega^2)^2 + k^2\Omega^2)}}$$

where

$$\alpha = \sqrt{(1 - \tfrac{1}{4}k^2)}, \quad \phi = \tan^{-1}(k/(1 - \Omega^2))$$

and A and B are constants.

As t becomes large, the first term on the right hand side becomes small and may therefore be neglected. As a result, the amplitude of the oscillations depends not on the initial amplitude but rather on the amplitude F and the frequency Ω of the driver. For a given value of F the amplitude of the solution is greatest when $\Omega^2 = 1 - \tfrac{1}{2}k^2$. If k is small, this amplitude can be very large compared with F; this is the well-known phenomenon of resonance.

We now turn to the simplest non-linear oscillator, the Duffing equation:

$$\ddot{x} + k\dot{x} + x + ax^3 = F \cos \Omega t, \qquad k > 0.$$

We write

$$\Omega = 1 + \omega$$

and we suppose that k, a, ω are all small. The cubic term represents the non-linear part of the restoring force. If $a > 0$, then the restoring force increases faster than linearly with the displacement and we have a 'hard spring', whereas if $a < 0$ we have a 'soft spring'. We have no term in x^2 because we want the restoring force to be an odd function of the displacement to make the motion symmetric about $x = 0$.

Since a is small, we expect that the solution of the Duffing equation will be close to that of the linear oscillator, so we attempt a solution of the form

$$x = A \cos(\Omega t - \phi).$$

We substitute this into the differential equation, ignore second order terms and a term in $\cos 3\Omega t$, and equate the coefficients of $\cos \Omega t$ and $\sin \Omega t$ on either side of the equation to obtain

$$\tan \phi = 4k/(3aA^2 - 8\omega)$$

and

$$A^2(\tfrac{3}{4}aA^2 - 2\omega)^2 = F^2 - k^2A^2.$$

The second of these is a cubic in A^2, so we recognize a cusp catastrophe with A^2 as state variable and a and ω as control variables. We locate the cusp point by differentiating the equation twice and eliminating A^2 from the two resulting equations; this gives

$$(a, \omega) = \pm (32k^3\sqrt{3}/27F^2, k\sqrt{3}/2)$$

so there are, in fact, *two* cusp catastrophes.

The situation is illustrated in Fig. 5.3. When $a = 0$ we have a linear oscillator, and there is a maximum of A at $\omega = 0$, which is the usual resonance effect that occurs when the frequency of the driver equals the natural frequency of the system. If we have a sufficiently hard spring, however, and if we begin with ω negative and then increase it, A will increase slowly to a maximum and then decrease, but it will then fall suddenly. At the same time there will be a change in the phase of the oscillator. If we now slowly decrease ω, there will eventually be another sudden change in amplitude and phase, but at a different point. A similar effect is observed in the case of a sufficiently soft spring. A full account has been given by Holmes & Rand (1976), who also carried out a

Fig. 5.3. The Duffing equation as a pair of cusp catastrophes. After Zeeman (1976*d*).

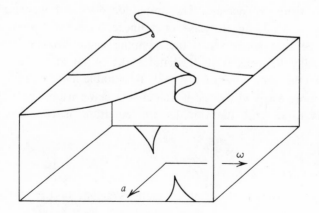

number of simulations on an analogue computer to verify that the assumption of a near-harmonic response is justified.

The Duffing equation has a number of applications in physics, but from our point of view its chief significance is that it illustrates another way in which catastrophe theory can apply to systems without gradient dynamics. It may be that a quantity which we can measure directly depends on the amplitude of some oscillation. If the governing equations are non-linear, then the quantity in question may exhibit discontinuous behaviour in accordance with the elementary catastrophes. In the present example we have a cusp – or rather, two cusps – but it is easily seen that if we add further odd powers of x to the restoring force we can obtain higher order cuspoids.

Not all oscillators which bifurcate do so according to the pattern of the elementary catastrophes, and so while we may expect to observe elementary catastrophes in systems with oscillators, there may also be bifurcations which are not on the list. Hence we are unlikely to be able to derive results like the one in the previous section about the naturally occurring caustics unless we happen to have additional information about the sorts of oscillators that are involved.

An example of an oscillator which bifurcates in a different way is given by the van der Pol equation,

$$\ddot{x} + k(x^2 - b)\dot{x} + x = 0, \qquad k > 0.$$

It is relatively easy to see how the solutions of this equation behave. When $b < 0$, the coefficient of $\dot{x}$ (i.e. the damping) is positive, and the origin is a stable equilibrium point. When $b > 0$, however, the situation is more complicated. For small values of x the damping is now negative, so the origin is an unstable equilibrium point and the amplitude of the oscillations tends to increase. The amplitude does not increase indefinitely, as it would in the case of a linear oscillator with negative damping, because when $x^2 > b$ the damping again becomes positive. What actually happens is that whatever the initial amplitude of the oscillations, the system eventually tends to a stable *limit cycle*.

This is more easily visualized if we define a new variable $y = \dot{x}$, which enables us to re-write the van der Pol equation as two first order equations:

$$\dot{x} = y,$$
$$\dot{y} = -k(x^2 - b)y - x.$$

Fig. 5.4 consists of three phase plane sketches of the system for different values of b and k. When b is negative, the trajectories spiral in towards the origin, but when b is positive, they tend to the limit cycle. When b is positive and k is large, then the limit cycle has the characteristic shape illustrated in Fig. 5.4c and the form of x itself resembles a square wave.

Although limit cycles are very important in many different fields, they tend not to be mentioned in elementary texts because they do not arise from linear equations. They bear a superficial resemblance to the solutions of the undamped linear oscillator in that they are periodic, but there are a number of significant differences. They typically occur as the solutions of structurally stable differential equations, and so we do expect to encounter them in nature. The amplitude of the oscillations depends

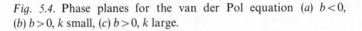

Fig. 5.4. Phase planes for the van der Pol equation (a) $b<0$, (b) $b>0$, k small, (c) $b>0$, k large.

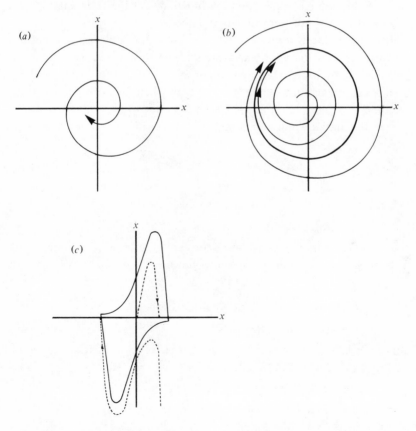

not on the initial conditions but on the equations themselves, i.e. on the structure of the oscillating system. By the same token, the solutions are stable; the system will return to the limit cycle after a perturbation, whereas a harmonic oscillator will continue in the new orbit to which the perturbation carried it. A limit cycle may be thought of as the natural generalization of an equilibrium point; it is an entire cycle of behaviour which is stable, rather than a particular set of values of the variables. It is likely that many biological phenomena, including the cell cycle, are controlled by limit cycles.

In all our previous examples, when a stable equilibrium point bifurcated it gave rise to an unstable equilibrium point with a stable one to either side. The van der Pol equation, on the other hand, undergoes a *Hopf bifurcation*, in which the unstable equilibrium point is surrounded by a stable limit cycle. The ideas developed earlier are not obviously relevant to this sort of behaviour; and in fact it can be shown that there is no stably bifurcating Liapounov function and that elementary catastrophe theory is not immediately applicable.

If the damping is large, however, we can still establish a link between the van der Pol oscillator and catastrophe theory. We can show this most easily by using a different phase plane from before; the trouble is that $\dot{x}$ is no longer a suitable variable with which to work because it can become very large (cf. Fig. 5.4c), so it is better to use $z = \int x$.

Let the initial values of x and $\dot{x}$ be x_0 and $\dot{x}_0$, respectively, and let

$$z(t) = z_0 - \frac{1}{K} \int_0^t x(\tau) d\tau,$$

where

$$z_0 = \tfrac{1}{3} x_0^3 - b x_0 - \dot{x}_0 / K.$$

We write the damping constant as the capital letter K to remind us that it is now taken to be large. We now have

$$\dot{z} = -x/K.$$

We substitute back into the original equation:

$$\ddot{x} + K(x^2 \dot{x} - b \dot{x} - \dot{z}) = 0,$$

and integrate, thus obtaining as an alternative form of the van der Pol equation

$$\dot{x} = -K(\tfrac{1}{3} x^3 - bx - z) \qquad \text{'fast equation'},$$
$$\dot{z} = -x/K \qquad\qquad\qquad \text{'slow equation'}.$$

We call these the 'fast' and 'slow' equations because with K very large, the rate of change of x is very much greater than that of z. Consequently z may be considered as a parameter for determining the behaviour of x. The equilibria of x are given by the equation

$$\tfrac{1}{3}x^3 - bx - z = 0,$$

and since we are treating z as a parameter we have a cusp catastrophe.

We interpret Fig. 5.5 in the following way. Off the surface, the fast equation ensures that the trajectories are very nearly parallel to the x-axis. The phase point will therefore move almost directly onto the surface. This makes $\dot{x}$ vanish, so the system is then governed entirely by the slow equation. If b is positive and constant, the system moves autonomously around an orbit (as shown), exhibiting the characteristic sudden jumps and hysteresis.

Thus a non-linear oscillator which does not bifurcate in accordance with elementary catastrophe theory can still, under certain circumstances, be represented as an elementary catastrophe with some additional structure, in this case a feedback flow (Zeeman, 1972b).

Fig. 5.5. The van der Pol equation as a cusp catastrophe.

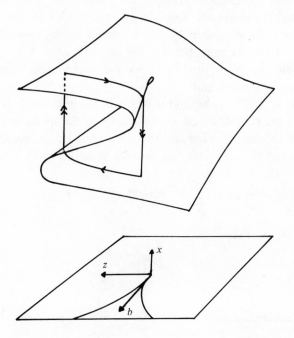

The decomposition of systems of differential equations into subsystems with different time scales is a device which is used in many other contexts (cf. Goodwin, 1963; Haken, 1977). It is sometimes called 'adiabatic elimination'. From a general point of view it is significant that the approximation which we had to make in order to apply catastrophe theory is the same as is sometimes necessary before other methods can be used.

The collapse of elastic structures

The study of the instabilities of elastic structures fits naturally into catastrophe theory because it is to a large extent based on the disappearance of local minima of a potential. Since it is generally possible to write down an expression for the potential energy of the system, the subject has not had to wait for catastrophe theory; the first results are due to Euler, who published them in 1744. Even the recent work of authors such as Thompson & Hunt (1973), who independently recognized some of the elementary catastrophes, was carried out from a classical point of view. In this section we use some of these results, originally obtained without catastrophe theory, as illustrations of the theory.

We begin with the simplest example, the Euler arch (Fig. 5.6). This consists of two light rigid rods of unit length, each with a supported free end. The other ends are joined by a pivot and a spring of modulus μ which tends to keep the angle between the two rods at π. A weight α is suspended from the pivot, and a horizontal compressive force β is applied to each of the free ends.

Let θ be the angle that either rod makes with the horizontal. Then the angle between the rods is $\pi - 2\theta$, so the energy in the spring is $\frac{1}{2}\mu(2\theta)^2$. The energy in the load is $\alpha\sin\theta$ and the work done by the compressive forces in buckling the arch into this position is $2\beta(1-\cos\theta)$. Hence the total energy of the system is

$$V(\theta) = 2\mu\theta^2 + \alpha \sin \theta - 2\beta(1 - \cos \theta).$$

Fig. 5.6. The Euler arch.

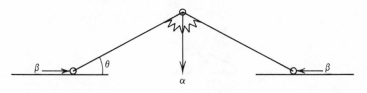

Then

$$V'(\theta) = 4\mu\theta + \alpha\cos\theta - 2\beta\sin\theta,$$
$$V''(\theta) = 4\mu - \alpha\sin\theta - 2\beta\cos\theta.$$

First we suppose that $\alpha = 0$, i.e. that there is no load. The condition for equilibrium is then

$$\beta\sin\theta = 2\mu\theta.$$

For $\beta < 2\mu$ this equation has only one solution, $\theta = 0$, and the corresponding equilibrium is stable. For $\beta > 2\mu$ the equilibrium at $\theta = 0$ is unstable, but there are two additional solutions, one with θ positive and one with θ negative, and these give stable equilibria.

This situation is familiar to us from our study of the catastrophe machines. In the present case it implies that if we apply a very small compressive force which we then increase, the rods will remain in a straight line for some time; only when β exceeds 2μ will buckling occur. This is indeed typical of the behaviour of many real structures: they can often withstand a considerable amount of stress (in this case compression) without any obvious response.

Let us suppose that the arch buckles upwards and that a load α is then applied. We can no longer handle the expressions in closed form, so we expand V in series:

$$V(\theta) \sim \alpha\theta - (\beta - 2\mu)\theta^2 - \tfrac{1}{6}\alpha\theta^3 + \tfrac{1}{12}\beta\theta^4.$$

Note that we are interested in small values of α and in values of β near 2μ, so it is the fourth order term which is the lowest that is bounded away from zero. We recognize V as the potential associated with the cusp catastrophe, and we put it into canonical form by transforming to a new variable

$$x = \theta - \alpha/2\beta.$$

Neglecting all terms of order greater than the first in α or $\mu - 2\beta$, and using $\beta \sim 2\mu$, we obtain

$$V(x) \sim \tfrac{1}{6}\mu x^4 - (\beta - 2\mu)x^2 + \alpha x.$$

We locate the bifurcation set by eliminating x from the equations $V' = V'' = 0$, and this enables us to predict that as the load α is increased the arch will gradually sag until the critical load

$$\alpha_c = \frac{10}{3\sqrt{\mu}}(\beta - 2\mu)^{3/2}$$

is reached. It will then snap suddenly into a downwards configuration.

This behaviour can be observed in real structures, although over-loading a bridge generally causes it to collapse altogether. This illustrates how we have to be careful when using equilibrium analyses, whether classical or within catastrophe theory. Once the arch starts to snap, it is no longer near equilibrium, and we cannot be sure that any further predictions are correct. Fortunately this is seldom a real drawback, as our main concern is to know if and when the equilibrium will become unstable; after that it is usually fairly obvious what will happen next.

We now consider a more realistic version of the same problem, the Euler strut (Fig. 5.7). The set-up is as before, except that the two rigid rods connected by a spring are replaced by a single flexible rod of length π (for computational convenience) and modulus of elasticity μ per unit length.

Let s measure length along the strut, and let $f(s)$ be the vertical displacement of the point s. We suppose that f and all its derivatives are small. The curvature of the strut is

$$\frac{f''}{(1+(f')^2)^{3/2}} \sim f'',$$

ignoring fourth order terms. The elastic potential energy of the strut is therefore

$$\frac{1}{2}\mu \int_0^\pi (f'')^2 ds.$$

The distance between the two ends of the strut is

$$\int_0^\pi (1+(f')^2)^{-1/2} ds \sim \pi - \frac{1}{2} \int_0^\pi (f')^2 ds,$$

so the energy lost by the compressive forces is

$$\frac{1}{2}\beta \int_0^\pi (f')^2 ds$$

and the total energy of the system is

$$V = \int_0^\pi F ds,$$

Fig. 5.7. The Euler strut.

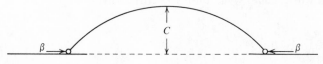

where

$$F \sim \tfrac{1}{2}[\mu(f'')^2 - \beta(f')^2].$$

We now have to find the configuration of the strut (i.e. the function f) which minimizes the total energy. Since the end points are fixed, this is a straightforward problem in the calculus of variations, and the condition for equilibrium is given by the Euler–Lagrange equations

$$(\partial F/\partial f'')' - (\partial F/\partial f')' = 0.$$

(See Thompson & Hunt, 1973, for a complete derivation.) Hence f is a solution of

$$\mu f^{(4)} + \beta f'' = 0,$$

and as it must also satisfy the boundary conditions

$$f(0) = f(\pi) = 0, \qquad f''(0) = f''(\pi) = 0,$$

with the latter pair expressing the condition that the strut is not curved at its ends, f is given by

$$f(s) = C \sin(s\sqrt{(\beta/\mu)}),$$

where C is a constant and $\sqrt{(\beta/\mu)}$ must be an integer. In particular, if $\beta < \mu$ then the only possible solution is $f(s) = 0$. Thus in this case too, below a certain value of the compressive forces no buckling occurs.

We can also show that if a load is applied to an upwards-buckled strut it will eventually snap into the downwards position, but since the calculation illustrates nothing that concerns us here, and since it will not be required later, we turn at once to a slightly different problem, the so-called 'pinned Euler strut'.

Imagine an Euler strut which has had its ends compressed until it buckled upwards, and suppose that the ends are pinned, as they would be in a real bridge. A vertical load α is then applied, but this time it is off-set from the mid-point by a small distance ε. We refer to ε as the imperfection, because it represents an inevitable manufacturing imperfection which produces a slightly asymmetric system. The boundary conditions on f are unchanged, and so without loss of generality we may write f in the form of a Fourier sine series

$$f(s) = \sum C_n \sin ns.$$

We recognize our previous solution as the first harmonic. We suppose (it can actually be shown to be true) that the chief effect of the asymmetric load is to bring the second harmonic into play, so we write

$$f(s) = x \sin s + y \sin 2s.$$

See Fig. 5.8.

At first glance this appears to imply that the state of the system is specified by two variables, and that consequently we are likely to encounter an umbilic catastrophe. In fact this is not the case. The distance between the ends of the strut is

$$d = \int_0^\pi (1 - (f')^2)^{1/2} ds$$

and since the ends are pinned, d is fixed, which gives us a relation between x and y. This relation cannot be expressed in closed form, but if we suppose that x is small and that y is of order x^2, and if we keep all terms as far as order x^6, then we obtain (Zeeman, 1976c)

$$d \sim \tfrac{1}{4}\pi(4 - x^2 - 4y^2 - \tfrac{3}{16}x^4 - 3x^2y^2 - \tfrac{5}{64}x^6).$$

Hence

$$x^2 + 4y^4 + \tfrac{3}{16}x^4 + 3x^2y^2 + \tfrac{5}{64}x^6 = r^2 + \tfrac{3}{16}r^4 + \tfrac{5}{64}r^6,$$

where r is the value of x when $y = 0$. We can readily verify that, to order x^6, this equation is satisfied by

$$x = r - \frac{2y^2}{r} - \frac{3ry^2}{4} - \frac{2y^4}{r^3}.$$

We are now able to derive an expression for the energy of the system as a function of y alone. Choosing units to make $\pi\mu = 1$ for convenience, we obtain for the energy in the strut

$$V_1 = \frac{1}{2\pi} \int_0^\pi (f'')^2 (1 + (f')^2)^{-3} ds$$

$$\sim \text{constant} + (3 + \tfrac{13}{8} r^2) y^2,$$

Fig. 5.8. The Euler strut with an asymmetric load. After Zeeman (1976c).

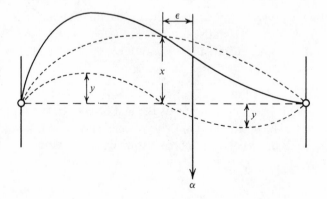

while the energy in the load is

$$V_2 = \alpha f(\tfrac{1}{2}\pi + \varepsilon)$$

$$\sim \alpha(x - 2\varepsilon y),$$

where we are still working to order x^6 and are also neglecting all terms of order ε^2. The total energy is therefore

$$V \sim \text{constant} - 2\alpha\varepsilon y + \left[\left(3 + \frac{13}{8}r^2\right) - \alpha\left(\frac{2}{r} + \frac{3r}{4}\right)\right]y^2 - \frac{2\alpha y^4}{r^3}.$$

The cusp is located at $\varepsilon = 0$, $\alpha = \alpha_0$, where

$$\alpha_0 = \left(3 + \frac{13}{8}r^2\right)\left(\frac{2}{r} + \frac{3r}{4}\right)^{-1}$$

$$\sim \tfrac{3}{2}r - \tfrac{1}{4}r^3,$$

so if we set $\alpha = \alpha_0 + a$ we may write

$$V \sim -\frac{3}{r^2}y^4 - 3r\varepsilon y - \frac{2}{r}ay^2.$$

Because the coefficient of y^4 is negative we have a *dual* cusp, with stable equilibria only on the middle sheet. Consequently as we increase the load α the equilibrium eventually becomes unstable. What happens after that cannot be predicted from our discussion, although Zeeman (1976c) has constructed a global model which enables him to show that the strut will snap into an asymmetric downwards configuration. Of course for many real structures this further analysis is of academic interest only, as the actual effect will be a collapse.

We remark that for a given strength of material, as measured by the modulus μ, it is the perfect (i.e. symmetric) structure which can support the greatest load. This is hardly surprising, but what we might not have expected is how rapidly the load-bearing capacity decreases with the imperfection (Fig. 5.9). Calculations carried out on a model of the perfect system will tend to overestimate the strength of any real structure by a significant amount, and this accounts for the interest that civil engineers have shown in these problems.

As a second example of the buckling of an elastic structure we consider the Augusti model. This is a less familiar system than the arch, but it provides a good illustration of the role catastrophe theory can play in these analyses.

We begin with the system shown in Fig. 5.10. This consists of a light, straight rigid rod of unit length, with a variable mass m at the top. The rod is pinned to a firm base and supported by a linear rotational spring of modulus μ. We let θ be the angle that the rod makes with the vertical. If the construction of the system were perfect, the spring would be unstrained when the rod was exactly vertical, but we introduce an imperfection by supposing that in fact the spring is unstrained when $\theta = \theta_0$, where θ_0 is small.

For θ not too large, the energy in the spring is $\frac{1}{2}\mu(\theta - \theta_0)^2$. The height of the point mass above the base is $\cos\theta$, so if we take the zero of potential energy to correspond to the vertical position of the rod we have

$$V(\theta) = \tfrac{1}{2}\mu(\theta - \theta_0)^2 + mg\cos\theta - \tfrac{1}{2}\mu\theta_0^2 - mg.$$

Fig. 5.9. Imperfection sensitivity of the Euler strut.

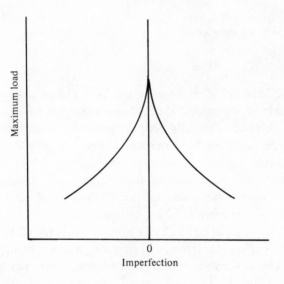

Fig. 5.10. The Augusti model.

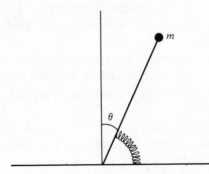

We now expand in series as far as the first term that is bounded away from zero, obtaining

$$V \sim \tfrac{1}{24}mg\theta^4 + \tfrac{1}{2}(\mu - mg)\theta^2 - \mu\theta_0\theta.$$

This is of course the potential for a cusp catastrophe with $\mu - mg$ and θ_0 as control variables, but the situation is not quite the same as for the Euler arch because we are not allowed to alter the control variables at will. Instead, we must choose θ_0 first and then leave it fixed, with only m allowed to vary. Consequently we do not need the full three-dimensional picture to allow us to visualize the behaviour of the system; it is sufficient to work with the projection onto the (m, θ) plane, with θ_0 as a parameter specifying which of the curves is appropriate.

Fig. 5.11 shows the equilibrium paths for the system. We can see that the perfect system remains vertical until $m = \mu/g$, and then gradually inclines to one side or the other. The imperfect system increases its original inclination as soon as any load is applied, with the rate of increase slow at first and then more rapid. But the system always remains stable; a small increase in the load always produces at most a small increase in the inclination. For this reason, Thompson & Hunt (1973) use

Fig. 5.11. Equilibrium paths for the simple Augusti model, showing the relation between the cusp surface usually sketched in catastrophe theory and the projection of it used in bifurcation theory.

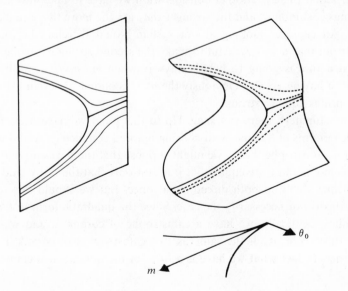

the term 'stable symmetric point of bifurcation' to describe what we call a cusp catastrophe.

The situation becomes much more interesting if we replace the pin at the bottom of the rod by a universal joint. The rod will now need two rotational springs to support it, so we supply a second at right angles to the first. The moduli of the springs are μ and v, and we let θ and ϕ measure the differences between the angles spanned by the springs and right angles. As before, we introduce imperfections by supposing that the springs are unstrained at $\theta = \theta_0$ and $\phi = \phi_0$. The potential energy of the system is

$$V = \tfrac{1}{2}\mu(\theta - \theta_0)^2 + \tfrac{1}{2}v(\phi - \phi_0)^2 + mg(1 - \sin^2\theta - \sin^2\phi)^{1/2}$$
$$- \tfrac{1}{2}\mu\theta_0^2 - \tfrac{1}{2}v\phi_0^2 - mg$$

or, expanding in series to the same order as before,

$$V \sim \tfrac{1}{24}mg\theta^4 + \tfrac{1}{2}(\mu - mg)\theta^2 - \mu\theta_0\theta$$
$$+ \tfrac{1}{24}mg\phi^4 + \tfrac{1}{2}(v - mg)\phi^2 - v\phi_0\phi - \tfrac{1}{4}mg\theta^2\phi^2.$$

Let us first consider the perfect system, with $\theta_0 = \phi_0 = 0$, and let us suppose that $\mu < v$. As m is increased, nothing happens until $mg = \mu$, at which point the coefficient of θ^2 becomes zero. The coefficient of ϕ^2 is still positive, so by the splitting lemma we can ignore ϕ, and it follows that the behaviour of the system is exactly as before. The rod simply inclines to one side or the other in the θ-direction.

The motivation for this example is, however, engineering, and this implies that there is another consideration we have to take into account. Springs cost money, and the stronger they are the more they are likely to cost. An engineer looking at our system would therefore be likely to point out that it was wasteful to make the second spring so strong. Since the structure is going to buckle anyway when $m = \hat{m} = \mu/g$, what is the point of having $v > \hat{m}g$? Obviously the most economical design is to have the springs equally strong.

Unfortunately, there is a snag. Up to this point we have been able to work in terms of θ alone, with ϕ being ignored completely. And if it were not for catastrophe theory we might suppose that there was no particular problem if μ and v are equal, that it meant only that the rod would begin to incline stably in both directions at once. But we should now realize that this is not necessarily the case. Since the quadratic form in θ and ϕ vanishes identically, we have a catastrophe of corank 2, and we know that this is not at all the same as two catastrophes of corank 1 taken together. In fact what we have is a form of the double cusp catastrophe which we met but did not study in detail in Chapter 3. We shall not attempt a full analysis here, either, but the following relatively simple calculation shows how different things are.

By setting $\phi = 0$ or $\theta = 0$ we can see that in the θ-direction or ϕ-direction the catastrophe is still a cusp, so the response of the system to a perturbation in either of these directions would be exactly as before, i.e. stable inclination. On the other hand, if we define new coordinates x, y by

$$\theta = x + y, \qquad \phi = x - y,$$

the potential of the perfect system becomes (after some algebra)

$$V(x, y) \sim -\tfrac{1}{6}mg(x^4 + y^4) + mgx^2y^2 + (\mu - mg)(x^2 + y^2)$$

so that in either the x-direction or y-direction the catastrophe is a *dual* cusp, or, in the terminology of Thompson & Hunt, an unstable symmetric point of bifurcation. Hence when the critical load is reached the system does not merely incline slightly; it fails completely.

This result may appear unexpected, but this is only because we did not carry our first analysis far enough. For the perfect system we have

$$\partial V/\partial \theta = \tfrac{1}{6}mg\theta^3 + (\mu - mg)\theta - \tfrac{1}{2}mg\theta\phi^2.$$

This implies (since ϕ is zero) that after the equilibrium at $\theta = 0$ becomes unstable, the inclination θ is related to the load m by

$$\theta^2 = 6(1 - \mu/mg).$$

The coefficient of ϕ^2 is therefore

$$\tfrac{1}{2}v - \tfrac{1}{2}mg - \tfrac{3}{2}(mg - \mu)$$

and a second bifurcation occurs when this vanishes, i.e. when

$$m = (3\mu + v)/4g.$$

It can be shown that at this point the system fails, although the instability is less obvious than in the case $\mu = v$.

There is thus nothing mysterious about the instability of the bifurcation in the optimized system; it is simply that the unstable second bifurcation has been made to coincide with the otherwise stable first one. Nor is it immediately obvious that this is a disadvantage. For let us suppose that we have a given amount of stiffness to apportion between the springs, so that

$$\mu + v = C,$$

where C is a constant. We write

$$\mu = \gamma C, \qquad v = (1 - \gamma)C.$$

Then the critical load at which failure occurs is

$$\hat{m} = (1 + 2\gamma)C/4g.$$

By hypothesis $\mu \leq v$, i.e. $\gamma \leq \frac{1}{2}$, and it follows that the maximum value of $\hat{m}$ corresponds to $\gamma = \frac{1}{2}$.

Thus, providing the system is perfect, the best design is the optimized one, $\mu = v$, whether we are concerned with the load the system can withstand without any buckling at all, or whether we are concerned with the load at which it fails. The only disadvantage is that the optimized system will fail without warning, whereas a non-optimized system will exhibit some stable buckling first. The snag, of course, is that it is not possible to build a perfect system, and it turns out that the imperfection sensitivity of the optimized system is much greater than that of the others. As a result, the actual gain in load-bearing capacity is considerably less than would be predicted from analysis of the perfect system, and may be insufficient compensation for the sudden collapses to which the system is prone.

Thompson & Hunt (1973) give a number of examples of similar effects. They point out that highly optimized designs are likely to have a number of undesirable properties. First, they are generally very sensitive to imperfections. Secondly, it may be very difficult to predict their behaviour under different circumstances. This arises from the fact that they lead to catastrophes of high codimension; for example even the most general form of the potential we wrote down for the Augusti model is not a versal unfolding of the double cusp. (This is the main point that catastrophe theory *per se* contributes to the discussion in this section, since the essentially classical analysis we have described does not warn us that more instabilities are possible. It appears, however (Poston & Stewart, 1978a), that the engineers have discovered the most important instability.) Finally, highly optimized structures are likely to have particularly nasty failure characteristics which, taken together with their high imperfection sensitivity, may lead to real disasters (one hesitates to say 'catastrophes') especially during construction.

6

Applications in the social sciences

The applications in this chapter represent the opposite end of the spectrum to the physical systems of Chapter 5. When we are trying to analyse the behaviour of an individual, or of a group, we cannot write down a set of equations of motion for the system based on known quantitative laws, and then look to see what catastrophe theory has to say about the solutions of these equations. What we must do is quite different. If we observe in a system some or all of the features which we recognize as characteristic of catastrophes – sudden jumps, hysteresis, bimodality, inaccessibility and divergence – we may suppose, at least as a working hypothesis, that the underlying dynamic is such that catastrophe theory applies. We then choose what appear to be appropriate state and control variables and attempt to fit a catastrophe model to the observations.

Right from the start, we see one of the advantages of catastrophe theory in this sort of problem. The data which are available are often not quantifiable. We can generally order our observations; for example, we can tell whether a person has become more angry or less angry. And we can usually say whether or not a variable is continuous and whether it changes smoothly. On the other hand, algebraic concepts such as addition and multiplication generally have no meaning: it makes little real sense to say that someone has become twice as angry. As a result, it is often very difficult to use differential equations or even many of the elementary statistical techniques in the social sciences. In catastrophe theory, on the other hand, the variables are defined only up to diffeomorphisms, so that order, proximity and smoothness are all we need. If we have data which are qualitative then we generally expect conclusions which are qualitative, and if we are going to go from one to the other by using a mathematical technique it seems natural that it too should be qualitative.

Conventions

There is a further point which we have to discuss before we can continue, and that is the question of the so-called conventions. Catastrophe theory tells us how many stable equilibria there are for a given choice of the control variables, but it does not tell us in which of them we will find the system. For the catastrophe machines or the buckling beams this is decided on what we might call historical grounds; the system remains in whichever equilibrium it happens to be until that equilibrium disappears. This behaviour is characteristic of most simple systems, but there are other possibilities, and we have to be prepared to deal with them.

Consider, for example, a gas which obeys the van der Waals equation of state,

$$(P + a/V^2)(V - b) = RT.$$

Here P is the pressure, V is the volume, T is the absolute temperature, R is the gas constant, and a and b are two constants characteristic of the particular gas. If we take the volume as the dependent variable, we may write this equation in the form

$$V^3 - (b + RT/P)V^2 + (a/P)V - ab/P = 0.$$

Since this is a cubic equation, the surface it represents is diffeomorphic to the equilibrium surface of the canonical cusp catastrophe. This suggests that we should be able to interpret the behaviour of the system in terms of catastrophe theory, taking V as the state variable and P and T as control variables.

If we attempt this, however, we discover that some of the predictions are wrong. It is certainly true that sudden jumps occur, since there is an abrupt increase in volume when a liquid changes to vapour. On the other hand, we do not observe hysteresis; water usually boils at the same temperature at which steam condenses. Nor is there bimodality, for we can predict the volume uniquely if we are given the temperature and the pressure.

The reason for the discrepancy is that the system does not remain in a local potential well until that minimum disappears. Instead, it seeks a global minimum of thermodynamic potential, with the result that the phase change occurs when the potential is the same for both phases. In terms of the diagram, the jumps in either direction occur as the control trajectory crosses a single curve which lies within the cusp (Fig. 6.1). The equilibrium surface does in fact resemble the simple P–V–T diagrams one finds in elementary texts on thermodynamics, except that it is on its side; the usual convention in thermodynamics is to take P upwards.

The criterion which determines which of two or more equilibria will be chosen is called a *convention*. This name is unfortunately somewhat misleading because the choice is not arbitrary; it is determined by the nature of the system. Systems which remain in the equilibrium that they are in until it disappears are said to obey the *perfect delay* convention. We expect to observe this in all simple systems because any other convention implies the existence of some means by which the system can discover another minimum and then move to it. Systems which always seek a global minimum of potential are said to follow the *Maxwell* convention; this name was chosen because it was Maxwell who formulated the rule which is used in thermodynamics to predict where the phase transition will occur. Other conventions are also possible: there is, for example, the *imperfect delay* convention, according to which the system will move to a global minimum, but only if this can be done without crossing a potential barrier of more than a certain height.

We encounter the Maxwell convention in thermodynamics because the process is governed by the laws of statistical mechanics, and there is

Fig. 6.1. The cusp catastrophe with the Maxwell convention.

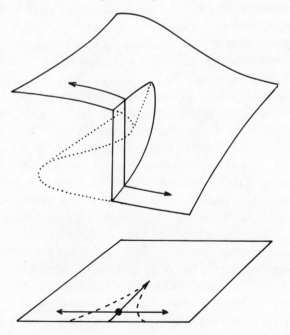

consequently an averaging involved. Intuitively we may say that the random motion of the particles in phase space allows the system to find the global minimum of potential. It is significant that it is possible, with sufficient care, to superheat water or supercool steam, although almost any disturbance will cause an immediate change of phase. What has happened is essentially that the reduced randomness is holding the particles more closely into the potential well, thus causing the system to follow the perfect delay convention.

Aggression in dogs

Our first example, one of the earliest and best known of Zeeman's many illustrations of catastrophe theory, is concerned with the behaviour of a dog under stress. In his book *On Aggression*, Lorenz (1966) claims that the two factors which chiefly influence aggression in dogs are rage and fear, and that these can be measured by direct observation of the animal; rage by how much the mouth is open and fear by how far the ears are laid back. It is clear enough that the level of aggression is increased by an increase in rage alone and decreased by an increase in fear alone, but what is the effect of a simultaneous increase in both factors? The answer appears to be that the dog will become either much more aggressive or much less aggressive, though it may be difficult to predict which. What is not likely is that it will remain calm, exhibiting the neutral behaviour of an untroubled animal.

We have here three of the characteristic features of the cusp catastrophe – bimodality, divergent behaviour and inaccessibility – so we try to fit a cusp to the observations. The control variables are clearly rage and fear, and the state variable is behaviour. Note that when two control variables are *conflicting* (by which we mean that increasing one of them alone generally has the opposite effect to increasing the other) the axes do not coincide with the u- and v-axes of the canonical cusp. Instead, we take the splitting variable to increase in the direction of the two factors increasing together. This corresponds to the idea that it is when the pressure is intense that smooth response can become impossible. The normal factor measures the balance between the conflicting factors.

We can now sketch Fig. 6.2 and see what it suggests about the dog's behaviour. We notice first that the picture reproduces the features we mentioned above: the clear effect of an increase of either rage or fear alone (paths 1 and 2) and the ambiguous effect of a simultaneous increase in both (paths 3 and 4). We also discover the two other phenomena typical of the cusp catastrophe: sudden jumps and hysteresis.

Fig. 6.2. The response of a dog to rage and fear (I).

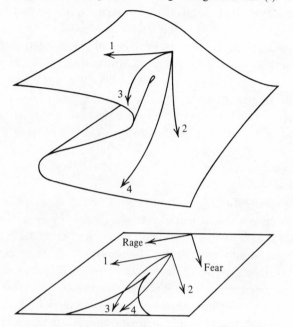

Fig. 6.3. The response of a dog to rage and fear (II).

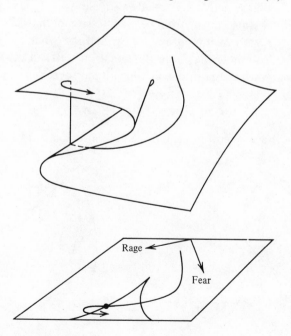

For example, if a dog is first frightened and then made angry (say by being confronted by a larger dog which then starts to invade its territory) it will not gradually become more aggressive. It is more likely to remain cowering (being trapped on the lower sheet of the model) and then exhibit an abrupt change to an aggressive posture. If this change brings it sufficiently far up the behaviour axis, the result will be an actual attack with practically no warning. Finally, once the dog is in an aggressive frame of mind, it will tend to remain so even if the rage is reduced (say by a partial retreat of the other dog) and so we observe hysteresis as well. See Fig. 6.3.

Decision making

We now turn from dogs to humans. For the sake of clarity we shall work in terms of one particular example, that of the government of a country which is in conflict with a rival nation, but the model is readily adaptable to other situations (Zeeman, 1976a, Isnard & Zeeman, 1976).

As control variables we take threat and cost. These are both as perceived by the government (since it is on a government's own estimates of them that it chooses its course of action) and both are to be understood as being defined in quite general terms, so that cost includes, for example, a loss of political stability at home, as well as the actual financial cost of military activity. The state variable is the type of policy to be adopted, on a scale ranging from very aggressive, or 'hawk', to very passive, or 'dove'. We can now draw a cusp and see what the model suggests (see Fig. 6.4); note that from this point on we shall draw only the control space, with the positions at which jumps occur indicated by dots.

Fig. 6.4. The response of a government to threat and cost.

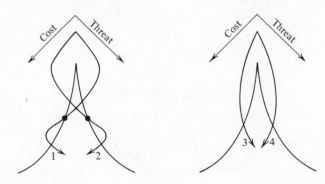

Path 1 represents a country which initially judges the threat to be increasing and the cost to be low. It therefore adopts an aggressive posture or, if a war is already in progress, will tend to escalate it. This policy will be continued even as the cost begins to rise. If the cost increases beyond a certain point, however, while the threat remains more or less the same, there will be a sudden switch to a much less aggressive position, and real attempts will be made to withdraw from the confrontation. An example of this behaviour is the history of the American involvement in Vietnam, including the failure to respond to the final offensive which resulted in the fall of Saigon.

Path 2, on the other hand, represents a country which initially perceives a high cost and a low threat. It therefore attempts to avoid becoming involved, and it will persist in this policy even if the threat is considerably increased. Should the threat become too great, however, there will be a sudden switch to a more aggressive policy, and this can even result in a quite unexpected declaration of war. It has been argued, for example, that a major factor contributing to the German decisions to enter Belgium in 1914 and Poland in 1939 was that in neither case had previous British actions given them reason to believe that Britain would enter the war or, once in, would continue to fight despite the terrible cost.

Paths 3 and 4 represent two countries which have reached more or less the same assessment of a situation, but which nevertheless adopt quite different policies. The reason for this divergent behaviour is that when cost and threat increase together there is a tendency to maintain the same sort of posture with which one began. An example of this could be the Cuban missile crisis of 1962; the Americans, who began more aggressively, kept the initiative throughout.

This model illustrates the diplomatic skill required in such cases. Had the Americans acted so as to increase the apparent threat to the Russians (say by threatening an invasion of Cuba) then the result might well have been disastrous, as the Russians could have been pushed over the edge of the equilibrium surface and onto the upper sheet.

Compromise

In the examples that we have considered so far, the response to a serious problem has been to move strongly in one direction or the other, with the middle ground barred by the inaccessibility characteristic of the cusp catastrophe. There are, however, examples of conflicts in which, despite the circumstances, compromises have been achieved. The cusp is evidently inadequate for a description of these, so we turn to the butterfly.

The butterfly catastrophe also has a single state variable and we again take this to be the type of policy, but there are now four control variables to interpret. As before, we choose threat and cost as conflicting variables in the normal-splitting plane. The compromise will be represented by points on the middle sheet, and since it is the butterfly factor which causes this sheet to appear we shall want to identify this factor with whatever we believe is chiefly responsible for making compromise possible. One plausible candidate is time: if a conflict continues long enough without either side winning, then there will eventually be a tendency for some sort of compromise to emerge. As for the bias factor, we can simply think of this as bias in the usual sense of the word, a tendency for a particular government to react to threat or cost more or less strongly than an alternative regime would. A change in the bias factor could then be brought about by a change in the composition of the government. One might prefer different choices for some of the variables – Isnard & Zeeman (1976) take the bias factor to be invulnerability, for example – but it turns out that our discussion is not affected by this.

Let us follow through a scenario based on Fig. 6.5. At the beginning

Fig. 6.5. The difficulty of achieving compromise.

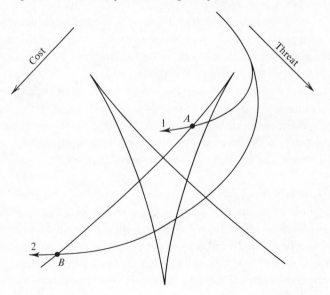

(not shown on the sketch) the butterfly factor is positive, so we have in effect a cusp catastrophe, and events will follow much the same course as in the previous model. We suppose that the issue is not decided one way or the other and that the causes of the tension remain. As time passes, the butterfly factor becomes negative (note that we have taken this factor as negative time to get the model the right way around) and the middle sheet appears. Providing the balance between threat and cost is not too much to either side, compromise is now possible. The middle sheet also becomes progressively wider, so that in time the compromise becomes less fragile, although there is always some risk that excessive changes in the other variables can destroy it.

Thus the two additional control variables of the butterfly catastrophe allow us to apply our model of decision making to more complicated situations, as indeed we would expect. The extended model is, however, more interesting in another respect, because in addition to providing a coherent picture of the process it also has at least one unexpected feature. The two paths marked in Fig. 6.5 are possible control trajectories for a country which initially adopts a hawk position but later finds the cost of the conflict to be increasing significantly. If the increase in cost occurs at a relatively early stage, then path 1 is followed, and at *A* there is a shift from the hawk position to compromise. This change of policy will not be very dramatic, as we are quite close to the region in which a continuous spectrum of policies is available. If the increase in cost occurs only after the threat has reached a much higher level, then the result, shown by path 2, is quite different. This time no appreciable change in policy occurs until *B* is reached, and then the hawk position is replaced not by compromise, but by the dove position. According to our model, once the situation becomes serious enough, the compromise position is inaccessible. It exists, but it cannot be reached. (See Fig. 4.10.)

This prediction is essentially due to the perfect delay convention, which holds the system in the hawk minimum while the compromise appears and disappears, so we are naturally led to ask whether or not this convention is appropriate. Let us consider what it means in terms of the system we are trying to model. We are studying a government which, we suppose, is trying to determine the best policy to follow, using its own estimates of threat and cost, and with its own biases and prejudices. (Note that the very use of the expression 'best policy' implies some sort of maximization principle, though we might be hard put to define precisely what it is that is being maximized.) The perfect delay convention states that a system will remain in a local minimum until that minimum disappears, even if a lower minimum exists. Translated into the language of the model, it says that a government will react to changes in the situation by only minor shifts in policy unless and until it finds its position totally untenable – even if by a more definite change of policy it could reach a position which *on its own criteria* is preferable.

Our model thus suggests that under certain circumstances a short-sighted government may fail to achieve a compromise which was available to wiser statesmen. This may not be a new prediction in the sense that no one has ever thought of it before, but it is certainly one that was not obviously built into the model.

There are (fortunately, if this model is correct) a number of alternatives to the perfect delay convention. One we have mentioned already, and this is the Maxwell convention, according to which a system will always seek a global minimum of potential. In our model this corresponds to a government with the imagination to be able to notice when a definite change of policy would be advantageous, and the confidence and political skill to be able to accomplish it. This is probably too much to hope for, and a more realistic choice of convention might be that of imperfect delay, in which the government may not adopt the better policy right away but at least does not wait until it has no alternative.

Another possibility is that governments, or individuals for that matter, are predisposed towards certain policies, so that they tend to adopt the one they prefer unless it is clearly untenable in the circumstances. This sort of behaviour, in which the choice between steady states is made according to criteria which are within the system but outside the model, is not to be expected in simple systems, but is more likely to be encountered in highly complex systems like the brain, which include numbers of interacting subsystems. An example might be the editorial policy of a newspaper at one end or the other of the political spectrum: very occasionally the paper may express a view sympathetic to the other side, but only in a situation in which no arguments whatsoever can be found in favour of the preferred position. And the return to the usual line occurs without hysteresis. (This convention has been proposed in a different context by Fowler (1972), who calls it the *saturation* convention.)

Finally, the choice of steady state can be made by some agent totally external to the system. Imagine a peace-keeping force which is sent to compel two warring factions to cease hostilities. The model predicts that providing the control trajectories are within the region under the middle sheet (i.e. providing neither side considers the balance between threat and cost to be too much to one side or the other), the arbiters may safely withdraw after a short time. If this is not the case, then the truce will break down as soon as they leave.

This 'external' convention can occur in other applications in the social sciences. For example, Zeeman (1976a, d) has suggested that catastrophe theory can be useful in the modelling of psychological disorders. In such cases the external convention explains the role of certain types of therapy, which can be seen as making it possible for the patient to reach a state which exists but which he cannot attain by his own unaided efforts. It also accounts for the fact that the therapy may work under certain conditions but not under others which may appear to be not very different; it depends on whether or not the desired equilibrium state is available.

Multistability in perception

An example of a switching phenomenon in psychology which is less dramatic than decisions about war and peace, but which is more easily studied, is multistability in perception. This is something which is familiar to anyone who has ever stared at a tiled floor, and the earliest account in the literature appears to be due to the geologist Necker, who reported in 1832 that line drawings of crystals appeared to reverse spontaneously in depth. The effect can be seen in Fig. 6.6; if we stare at the corner marked A for some time it will alternate between appearing to face forward and appearing to face back. Another simple example is

Fig. 6.6. The Necker cube.

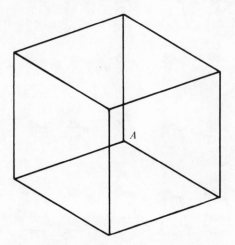

shown in Fig. 6.7; this is tristable, because we can see either the left hand shape, or the right hand shape, or else the line between the two as a thing in itself, rather than as a boundary.

An interesting extension of the phenomenon, due to Fisher (1967), is illustrated by the top row of figures in Fig. 6.8. Here there is a gradual

Fig. 6.7. A tristable figure.

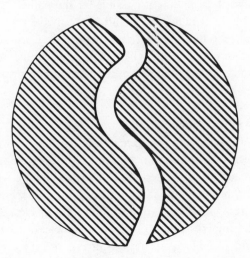

Fig. 6.8. Bistability as a cusp catastrophe. From Poston & Stewart (1978a, p. 419).

shift from what is clearly the face of a man on the left to what is equally clearly a woman on the right, with the fourth figure from the left having equal probability of being interpreted as either. What is especially interesting from our point of view is that if we view the figures in sequence, then those in the middle are likely to be seen as a man if we begin from the left but as a woman if we begin from the right.

We have three of the characteristic features of the cusp catastrophe – bimodality, sudden jumps and hysteresis – so we may suspect the existence of a cusp. Of course the top row by itself is only a one-dimensional range and could also be fitted by two folds (Fig. 6.9) but we prefer the cusp (i.e. we investigate this possibility first) because catastrophe theory tells us that it is the simplest way to reproduce the observations with a single organizing centre. Poston & Stewart (1978a, b) have suggested detail as an appropriate choice for the second control variable, and we can see the effect of this in the complete Fig. 6.8. We can pass from what is definitely a man to what is definitely a woman without a sudden jump if we reduce the detail to the point where there is insufficient evidence to allow us to decide one way or the other, and we can construct sequences from one of the middle figures on the bottom row to one of the middle figures on the top row which will bias the interpretation strongly in favour of either man or woman.

One feature which the model does not include is the alternation of the Necker cube, the switching between two steady states with no change in the control variables. Neither does it account for the phenomenon of

Fig. 6.9.

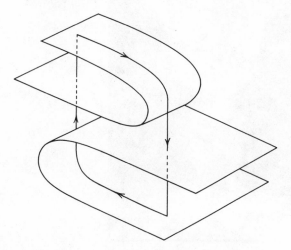

suggestion. In many ambiguous figures we can see only one of the two interpretations until the other is pointed out to us, though after that they may appear to alternate. Even in the case of the Fisher figures we can induce a switch before the cusp is crossed by suggesting the alternative to the observer.

We can fit these into the model if we suppose that the external convention can apply, i.e. that in complex systems such as the brain the mechanism which determines the possible steady states need not be the same as the mechanism that chooses among them. In the case of ambiguous figures it is as though within a certain range that part of the brain which normally decodes what we see is unable to reach a firm conclusion and passes two alternatives along for a decision on other grounds; the well-known goblet-and-faces figure of Rubin (Fig. 6.10) is much more likely to be seen as a goblet in a book on glassware. Perhaps decision making is somewhat similar; we may find ourselves in a position in which we can rapidly narrow down the alternatives from the very large number which are in principle available, but something different may be required to choose from among the few that remain.

More on conventions

We have been led to the 'saturation' and 'external' conventions by our attempts to apply catastrophe theory to the social sciences, but

Fig. 6.10.

they can also be observed in physical systems. Consider, for example, a Zeeman catastrophe machine (Fig. 1.1) and suppose that the free end of the elastic is level with the curvilinear diamond but to the right of it. The pointer OQ can be at equilibrium only if it is inclined to the right. We can move it to a leftwards position, but it will always return to the right.

Now move the free end P towards the left. When it enters the diamond, nothing dramatic happens spontaneously; the pointer remains inclined to the right. But there has been a change in the system, for if we now move the pointer to the left it will stay there. Equally, so long as P is within the diamond we may move the pointer back and forth at will between the two stable orientations, but if P is outside the diamond then there is a unique equilibrium position to which OQ will always return.

More interesting illustrations of the same effect arise from the Duffing equation. For example, if we stop a vibrating hard or soft spring by pinching it between our fingers while the driver is still running, we would normally expect that when we released the spring it would, after a transient phase, return to the same amplitude of oscillation that it had before. If, however, the control trajectory was within the cusp, and if the spring happened to be in the mode corresponding to the larger amplitude, then the subsequent oscillations might well stabilize at the smaller amplitude. And with care we could shift the system from small to large amplitude oscillations, as well.

Another realization of the Duffing equation can be obtained by constructing a suitable electrical circuit. In this case the amplitude could be changed by the introduction of a pulse of current sufficient to drive the amplitude away from one attractor and towards the other. Not only could this effect be used as a demonstration of the cusp catastrophe, it might also provide an explanation of some of the phenomena we have been discussing in this chapter. It has been known for a long time that there are electrical oscillations in the brain, and Zeeman (1976*d*, *e*) has proposed that the brain can be modelled as a large collection of coupled oscillators. Abrupt changes in mood or behaviour can then be understood as due to catastrophic jumps in the amplitude and phase of non-linear oscillations. The external convention fits into this model quite naturally, if we suppose that the effect of a stimulus such as suggestion is an electrical pulse imposed on the relevant oscillating circuit. This can result in a change if, but only if, the other attractor exists.

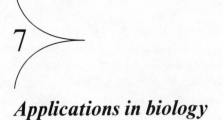

7

Applications in biology

As one might expect, the applications of catastrophe theory in biology tend to occupy a position on the spectrum somewhere between those in physics and those in the social sciences. We do not usually know the dynamic, but we do generally have at least some idea of the processes involved. As a result, we are often in a position to judge whether or not the conditions necessary for catastrophe theory to be applicable are likely to be satisfied, and this puts us on much firmer ground than in the social sciences. Indeed, one of the aims of applying catastrophe theory in biology is to help us in the task of deducing the mechanism.

The two examples that we discuss in this chapter differ from those in Chapters 5 and 6 in an important respect. So far we have seen catastrophe theory applied to problems which had previously been studied by other methods. Here, in contrast, we have two case studies of catastrophe theory in action. In both cases, new results were obtained and (which should satisfy those who see it as the sole criterion for the usefulness of theory in science) further experiments were suggested.

The movement of a frontier

This is one of the first real applications of catastrophe theory. We begin with the statement and proof in more or less the same form as that originally given by Zeeman (1974). We then discuss part of the rest of that paper in order to see precisely what it is that the analysis accomplishes.

We want to show the following: when a frontier forms in a previously undifferentiated region, the four hypotheses

1 homeostasis
2 continuity
3 differentiation
4 repeatability

imply that the frontier generally does not first appear in its final position. Instead it forms somewhere else, moves as a wave through the region, and then stabilizes and deepens. The final position is reached parabolically, not approached asymptotically. Mathematical interpretations of the hypotheses are given in the proof; in any application they have to be given physical meanings which are appropriate to the system under consideration and consistent with the precise formulation. In this example we are primarily concerned with embryology, so the region is part of a tissue and we use the following definitions.

Homeostasis: Each cell is in a stable biochemical equilibrium, though this equilibrium may change with time.

Continuity: At the beginning of the experiment, the chemical, physical and dynamical conditions in different cells can be represented by a smooth function of position within the tissue. This continuity implies that neighbouring cells will follow nearby paths of development wherever possible.

Differentiation: Whereas at the beginning of the experiment there is only one type of cell with at most a smooth variation in properties, at the end of the experiment there are two distinct types, with a sharp frontier between them.

Repeatability: The process is structurally stable.

The first step in the derivation is to remark that even though the region in which the frontier forms will generally be two-dimensional or three-dimensional, we are really concerned only with the one-dimensional interval which is orthogonal to the frontier. If we denote this interval by S and let T be a time interval which encompasses the development, then $C = S \times T$ is the rectangle of space–time which includes the entire process and is therefore the natural choice for the control space. We now suppose that n state variables are required to specify completely the state of a cell, and this makes $R^n \times C$ the phase space for the process.

We have already assumed that each cell is in stable biochemical equilibrium, and we now interpret this to mean that the state of each cell corresponding to any $(s, t) \in C$ is some $\mathbf{x} \in R^n$ which can be found by locating the minima of a potential or, more generally. Liapounov function, $V(\mathbf{x}, s, t)$. We take this function to be smooth and generic. Our hypotheses now permit us to apply catastrophe theory, and we take as the essential state variable x the quantity in which the discontinuity appears. We shall refer to this as the morphogen.

Fig. 7.1 illustrates an attempt to sketch the stationary values of $V(x, s, t)$. We know that the values are continuous along the curve *ab*, on account of hypothesis 2, and that there is a discontinuity along *AB*, on account of hypothesis 3. The problem is to complete the picture with one of the surfaces we drew in Chapter 4. The simplest catastrophe is the fold, but this cannot be made to fit, so we try a cusp.

It is clear that if we draw a cusp opening to the right we can complete the diagram, but there are two further points to consider. First, we want the potential to be generic, so we must not choose the axis of the cusp to lie precisely in the *t*-direction. (See, however, the discussion at the end of this section.) Moreover, the cusp will have to curve back because, as we shall see, this is necessary if the frontier is to stabilize and not move on to the far end of the tissue. This non-local requirement does not affect the issue of whether or not the frontier moves, but if it is not satisfied then at the end of the experiment there will be no frontier. Hence we choose Fig. 7.2*b* in preference to the simpler Fig. 7.2*a*.

We now have a model (Fig. 7.3) which is the simplest that will suffice. Let us investigate its properties. A line drawn on the surface $\nabla_x V = 0$ and parallel to the *t*-axis traces out the history of one particular cell

Fig. 7.1.

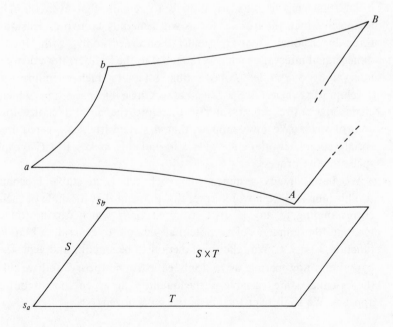

throughout the experiment; it is the 'world line' of that cell. A line drawn parallel to the *s*-axis corresponds to a snapshot of the whole tissue taken at a fixed point in time. By following some of these lines we can deduce what happens as differentiation proceeds.

Any cell in the interval $s_a < s < s_1$ develops smoothly into an *A* cell, and any cell in the interval $s_2 < s < s_b$ develops smoothly into a *B* cell. The behaviour of cells in the interval $s_1 < s < s_2$ is different. They develop smoothly until their control trajectories cross the cusp for the second

Fig. 7.2. Two attempts at completing Fig. 7.1: (*a*) non-generic, (*b*) generic.

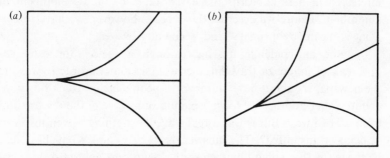

Fig. 7.3.

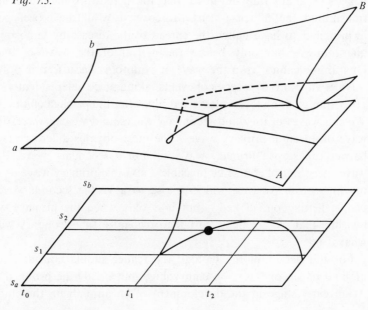

time, at which point there is a sudden change in the concentration of the morphogen x. They then develop smoothly into A cells. If it is possible to monitor the concentration of x throughout the tissue, it will be seen to vary continuously from one end to the other until time t_1. At this time a distinct frontier will appear, although at first the discontinuity in morphogen concentration will be very small. The frontier will move through the tissue until time t_2, when it will have reached its final position, s_2. It will then remain stationary, although it will probably deepen further. This establishes the result; we now proceed to see how it is applied.

The paper in which this result first appeared was entitled 'Primary and Secondary Waves in Developmental Biology', and the part we are interested in here concerns the early stages of the development of an amphibian. Before we describe the process, however, we have to explain what is meant by 'primary' and 'secondary' waves.

Imagine an epidemic moving through a region, for example, an outbreak of influenza travelling across Asia and Europe. By reading in the newspapers about large numbers of people being struck down in one country after another we form a picture of a wave of the disease moving from east to west. But is this a real wave; is anything – even just a signal – being transmitted? The answer is clearly no. What has actually happened is that some time before the symptoms appeared there passed through each country a wave of infection. This wave was invisible (to the naked eye, at any rate) but it was real, for there was something real being transmitted, the influenza virus. Moreover, it would have been possible, in principle, to have halted the spread of the disease by imposing strict quarantine – but only before the arrival of the wave of infection. Quarantine cannot stop the wave of symptoms because it is not a real wave at all, but only a series of events taking place independently of each other though at a more or less fixed time after the passage of a real wave. We call waves of this kind secondary waves, and we call waves like the wave of infection primary waves. The most obvious difference is that a barrier can stop a primary wave but not a secondary wave. Primary waves need not of course be invisible – a visible primary wave of sunrise moving across America will give rise to a visible secondary wave of people getting out of bed – but it is only when the primary wave is invisible that we are likely to fail to recognize the secondary wave for what it is.

For full details of the biology the reader should consult Zeeman's original paper, or a text on embryology, but briefly the problem is this: At an early stage in the development of an amphibian, the embryo is approximately spherical and partly hollow. An opening called the blastopore then appears, and the mesodermal tissue, which is on the outside of the sphere, flows through the blastopore and takes up a position on the inside. This involves an active process which Zeeman calls 'submerging': the cells decrease their free surface and increase the proportion in contact with other cells. This causes the mean curvature of the tissue to change from positive to negative, which is necessary because what was formerly the convex side of the layer has to become the concave side.

The submerging is observed as a wave passing through the tissue; the point at which it occurs is fixed in space but since the tissue as a whole is moving, the cells submerge in sequence. The problem is to explain the origin of this wave. If it is a primary wave, then it needs its own control mechanism, some sort of signal passing through the tissue. If, on the other hand, it is a secondary wave, then everything is much simpler. All that is required is some mechanism to make submerging take place at a specified time after the passage of the primary wave, and we can imagine a number of ways in which this might be arranged.

The simplification only works, however, if we can find the primary wave. Moreover, the primary wave must have some other function as well; if its only purpose is to trigger the submerging we are no further ahead. So we look back at the events which take place in the embryo shortly before gastrulation. We find nothing that is obviously wave-like, but we do notice that at an earlier stage the entire upper portion of the embryo was undifferentiated. It only later differentiated into mesoderm and ectoderm. The result we have just established now tells us that this frontier is likely to have moved as it formed, and this could be the primary wave we are looking for.

It would not be correct to say that we have proved that the process takes place in the way we have described. What we have shown, however, is that there is a strong likelihood that two apparently unrelated events are in fact closely connected, and are initiated by the same stimulus. Rather than trying to discover the signal that causes the cells to submerge, we are led to look for the hidden primary wave of differentiation, and indeed there is now experimental evidence for its existence (Elsdale, Pearson & Whitehead, 1976; see also Zeeman, 1978). This is not really very different from what happens when we use traditional mathematical techniques; the theory may tell us what to expect and, if we are fortunate, it may even predict effects which our unaided intuition would have failed to anticipate, but ultimately it is only through observation and experiment that we can know whether we are right.

If it is typical for a wave to be associated with the formation of a frontier, then we may expect waves to be relatively common in development, and this may be the clue to the elucidation of a number of processes which otherwise appear difficult to explain. An example of this is the formation of somites, segments which appear during the early stages of the development of vertebrates. Most models of this process involve some sort of 'prepattern': an essentially static chemical pattern is established first, and this serves as a template for the somites. Cooke & Zeeman (1976) have now proposed a model in which the primary wave interacts with an oscillator in each cell. This model appears to be in good accord with the observations and has the advantage that it is able to account for the phenomenon of regulation (the ability of the embryo to produce the correct number of somites over a wide range of body sizes and in spite of considerable variations in the environment).

Once we suspect that frontiers often move as they form, we can think of mechanisms to account for this. For example, if we suppose that the morphogen is produced within each cell separately but that the reactions are all moderated by some substance which can diffuse through cell walls, then we are led to the class of reaction–diffusion equations, and these are well known to have solutions which are travelling waves. Moreover, it has already been suggested that such equations may be involved in development. The reason for this is that the coordination of certain processes is very difficult to explain if communication is by chemical diffusion alone, as this is very slow. Reaction–diffusion waves travel much more rapidly through tissue. Thus Zeeman's work indicates that a class of reactions which are already considered to be of importance in development may be playing a further, and hitherto unsuspected, role.

There may well come a time when we are able to write down and solve the equations that govern the concentration of the morphogen. The analysis based on catastrophe theory has removed the need for us to wait for that to happen before we can use the motion of the frontier in our attempts to understand the very difficult problem of development. And it was this analysis which led experimenters to look for the wave, and which may provide the clues needed to elucidate the mechanism.

There are, of course, other situations in which frontiers form in regions which were originally undifferentiated, and since the arguments we have used to derive the result do not depend on the precise details of the application we may expect them to be valid in at least some of these other cases as well. Consider, for example, the edge of a forest. There is usually quite a sharp division between the woods on one side and the grassland on the other. It is difficult to account for this solely in terms of environmental variations, as these are generally much more gradual, and would lead us to expect a transition zone. If such a zone existed, however, there could well be competition within it; grass tends to choke seedlings, but where trees are able to develop they can eliminate grass by shading it. This process can presumably be modelled by differential equations, and then suitable ecological interpretations of the four hypotheses lead us to the conclusion that this frontier too should move as it forms (Zeeman, 1974). There is now in the literature (Ashton, cited by Poston & Stewart, 1978*a*) a detailed study of such a situation in which it is shown that not only did the frontier move, but that when it eventually stopped it did so parabolically, rather than exponentially.

It is interesting to consider what might have happened without catastrophe theory. We can imagine that eventually it would be noticed that the frontier between ectoderm and mesoderm moves as it forms, and that the frontier between woods and grass also moves. In time, other examples would be found. After considerable effort, theoreticians in the different fields would write down differential equations which would, for each case separately and with varying degrees of accuracy, describe the differentiation and account for the motion of the frontier. Eventually someone might notice that the same phenomenon was occurring in several different contexts, and he would attempt to determine whether or not there was some common underlying explanation. In the end, he might well conclude that there is a large class of differential equations which lead to the formation of frontiers in previously undifferentiated regions, and that the solutions of these equations typically predict motion of the frontier.

Catastrophe theory allows us to approach the problem from the other end. If a certain effect is common to a large class of differential equations then (provided that it is the sort of phenomenon that catastrophe theory deals with) we are likely to realize this at once. We are then led to look for it in various individual cases, including some in which we would be very unlikely to notice it if we did not already suspect its existence.

When Zeeman's paper was first published, it provoked a certain amount of controversy, most of which centred around the question of the orientation of the cusp. Since this step is crucial to the result, and since it does not seem to follow in a natural way from the hypotheses, it is instructive to consider the point carefully.

Following Zeeman, we have argued that taking the potential $V(x, s, t)$ to be generic means that the axis of the cusp cannot be taken parallel to the t-axis. This is a perfectly standard line of reasoning, and one which has been used in proving some of the mathematical theorems which are not in dispute, but we have to be careful to check that the genericity that is implied is appropriate to the model. There can be symmetries in a system which prevent some of the control variables from being altered, and there is also the question of (r, s) stability (Wasserman, 1976) in which we have to allow for the possibility that not all the control variables can be mixed freely. Space and time are not always interchangeable, after all.

It might be an interesting exercise to see if these considerations are relevant to the present problem and, if so, what additional hypotheses are required. We shall not deal with these matters here, however, because there is a much simpler objection, viz. that there is nothing in our result – even if we succeed in tidying it up – that requires the frontier to move by more than a tiny fraction of the width of a single cell, and in that case no motion whatsoever would actually be observed. Moreover, no appeal to genericity can make the cusp curve back, and we have seen that this is essential if the frontier is to stop.

The solution to all these difficulties is to recognize that in order to apply the result it is not necessary to prove that all frontiers must move, much less that they must all move more than some infinitesimal distance. All that is needed is to establish that among the simplest plausible mechanisms which can result in the formation of a frontier, a large class have the property that they cause the frontier to move as it forms. This we have done. The motion of a forming frontier thus becomes something to be expected, rather than the highly unusual phenomenon our intuition might lead us to imagine it to be. We see here an example of how in using catastrophe theory it is essential to keep in mind exactly what it is that we are trying to understand.

Determination of critical variables

Our second example (Bazin & Saunders, 1978) arose out of what began as an attempt to create a reliable system for studying predator–prey dynamics but which, chiefly on account of an analysis based on catastrophe theory, led in quite a different direction.

The mathematical study of the interaction between a predator and its prey goes back to the 1920s, when an Italian biologist, d'Ancona, noticed that there were differences from year to year in the total numbers of two species of fish caught in the Adriatic Sea. The variations did not appear to be related in an obvious way to changes in the environment because the numbers did not go up and down together. Instead, the fluctuations in the population of the larger species seemed to follow those of the smaller. It was known that the larger species preyed on the smaller one, and this suggested to the mathematician Volterra the following simple model, which has become known as the 'Lotka–Volterra equations':

$$\dot{H} = \alpha H - \mu H P,$$
$$\dot{P} = \nu H P - \beta P.$$

Here H is the number of prey organisms, P is the number of predators, the Greek letters all denote constants, and a dot indicates differentiation with respect to time. We are supposing that in the absence of the predators the prey will increase exponentially, that in the absence of prey the predators will decrease exponentially, and that both the rate of feeding and the rate of growth of the predators are proportional to the product of the two populations. The equations cannot be solved in closed form, but it can be shown that the solutions are periodic, with P reaching its maximum value about a quarter period after H.

Despite this apparent success, there are reasons for doubting that the Lotka–Volterra equations provide a good explanation for the population oscillations that are observed in nature. Clearly they are at best only a very rough approximation to the actual interactions between species, and since the equations are also structurally unstable this casts doubt on any conclusions we may draw from them. Quite apart from such mathematical considerations, there is the difficulty that there does not appear to be in the literature a really convincing example of oscillations which originate in the interaction between a predator and its prey. When oscillations have been analysed in detail they have generally been shown to be driven by variations in the environment or, less commonly, to be limit cycles arising out of more complicated interactions.

Now at least part of the reason that there are few if any examples of population oscillations of the Lotka–Volterra type may simply be that it is difficult to obtain data over a sufficiently long period of time and sufficiently free from the effects of sampling error and environmental perturbations. So it was decided to set up in the laboratory a system consisting of two species of microorganisms, which reproduce very quickly and which can be grown in very large numbers. The organisms, an amoeba, *Dictyostelium discoideum*, preying on a bacterium, *Escherichia coli*, were grown in a continuous culture vessel called a chemostat. Medium containing nutrient (for the bacteria) entered through one port, while medium from the vessel was drawn off through another. This system has many technical advantages, but for the present discussion the chief significance is that there was a continuous loss of both species of organisms from the chemostat at a rate directly proportional to their populations. The constant of proportionality is called the dilution rate.

A typical set of results is shown in Figs. 7.4 and 7.5, and it is clear from these that the system was behaving in a manner which is not easily accounted for by a model based on the Lotka–Volterra equations or some simple modification of them. The most obvious peculiarity is that the plot of P against time consists essentially of a succession of straight line segments. Now the slope of this curve measures the excess of the amoebal specific growth rate, λ, over the (constant) dilution rate. It follows that λ itself was remaining constant over quite long periods of time (especially when compared with the generation time of these organisms, which is only about 3 h) during which the prey density could change by a factor of as much as 100. It would then change abruptly to a

Fig. 7.4. Prey biovolume (data from Dent *et al.*, 1976). After Bazin & Saunders (1978).

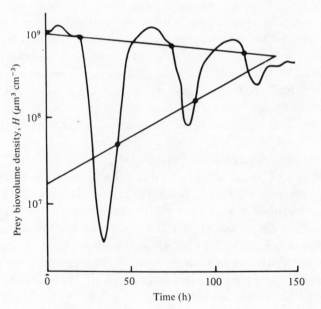

different value, which would again be maintained for a considerable time, despite large changes in the prey density.

The occurrence of sudden jumps suggests that catastrophe theory might provide a suitable tool for analysis, and the first thing to do is to choose a suitable catastrophe and fit the observations to it. As in the previous example, we find that the simplest catastrophe that will do is the cusp, because the fold cannot reproduce the feature that as the experiment continued the jumps became progressively smaller and eventually disappeared altogether. The state variable is of course λ, since it is in this quantity that the discontinuities were observed. An obvious choice for one of the control variables is the prey biomass H, because it is generally assumed that this is the factor which chiefly determines λ. In fact, most studies assume that λ is governed by H alone, but we need a second control variable, and we choose the time, t. This seems a natural enough choice, especially as there appear to have been time-dependent changes in the system (not shown on the figures here; see Dent, Bazin & Saunders, 1976 for details), but it does have a significant consequence: if we succeed in modelling the system we will not yet know the differential

Fig. 7.5. Predator biovolume (data from Dent *et al.*, 1976). After Bazin & Saunders (1978).

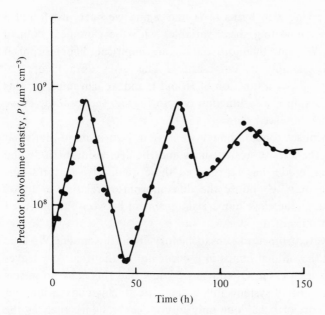
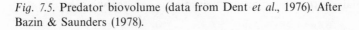

equations which govern it, but we will know that they are not any of those customarily used in predator–prey studies, as these are autonomous.

With this choice of variables, Fig. 6.4 becomes a drawing of the control space for the system, with the control trajectory indicated. We now mark the points at which the sudden jumps in λ occur and sketch a cusp-like curve connecting the jumps up on one branch and those down on the other; we see that a pair of straight lines will give an adequate fit. The next step is to construct a diffeomorphism from the control variables H, t to the canonical control variables u, v which will carry the pair of lines into the curve $27v^2 = 8u^3$. We do this by translating the origin to the point of intersection of the lines, rotating the axes so that the new coordinate axes are the bisectors of the angles between the lines, and then making a further transformation of the form

$$u \mapsto au, \qquad v \mapsto bv^{2/3},$$

where a and b are constants. This diffeomorphism appears to be the simplest that will accomplish what we require, but it is by no means unique.

For the canonical cusp catastrophe, the state variable x can be obtained from the control variables u, v by solving the equation

$$4x^3 + 2ux + v = 0.$$

We have data giving λ in terms of H and t, and we have just found a diffeomorphism connecting these variables with the canonical control variables u, v. We can therefore obtain an empirical diffeomorphism relating x and λ simply by curve-fitting. This gives us a theoretical relation expressing λ as a function of H and t, and we can integrate this numerically to obtain P as a function of t. In Fig. 7.5 the resulting curve is compared with observed data.

The fit is obviously good, although we have to bear in mind that what is significant is the almost straight lines and the decreasing periods and slopes. Once the model has reproduced these qualitative features correctly, we have only to choose the diffeomorphism relating x and λ carefully to arrange the close numerical agreement between the observed and predicted values of the slopes.

There is, however, something unsatisfactory about this model. We note that in Fig. 7.4 the sudden jumps in λ occur not as the trajectory leaves the cusp, but as it enters it. Of course we know that this is by no means impossible, since not all systems follow the perfect delay convention, but we expect this sort of behaviour only in complex systems, such as the brain. In those it is not implausible to imagine some separate switch which acts to select a particular steady state as soon as it appears, even if this involves moving to a higher potential, but in our example such an explanation seems less likely. One could suppose that the amoebae can sense the rate of change of bacterial biomass density and that they try to grow either rapidly or slowly, depending on the sign of this derivative. If the governing biochemical equations are such that only within the cusp are both the rapid and slow growth modes available, this will produce the observed behaviour. It does, however, seem odd that such relatively simple organisms as amoebae should be using two separate switches to govern a single and basic response.

In conventional modelling we sometimes find that a model which we believe to be along the right lines nevertheless leads to an unsatisfactory conclusion. When this happens, we generally try to discover some modification which will remove the anomaly without forcing us to abandon the model altogether. Quite frequently we succeed in this, and when we do we generally find that we have learned something more about the system we are studying. We can attempt to do the same sort of thing when using catastrophe theory, although because we make so few assumptions about the system we have very little room for manoeuvre. The only real freedom we had in constructing the model (as distinct from choosing the diffeomorphisms in the final curve-fitting stage) was in the choice of control variables, so before either rejecting the model as implausible, or else setting out to elucidate the rather complicated mechanism which appears to be involved, we ought to see if we could have made a better choice.

Most models of microbial growth do, as we said earlier, assume that the specific growth rate of an organism is chiefly determined by the concentration of its limiting nutrient. There is, however, some evidence which suggests that for amoebae the correct critical variable is the specific concentration of nutrient, i.e. the ratio of bacterial biomass to amoebal biomass, or H/P in our notation. To see the effect of this hypothesis we plot H/P as a function of t, and mark on this diagram the locations of the sudden changes in λ. The result (Fig. 7.6) is striking. The undesirable feature of Fig. 7.4 has disappeared; the configuration of jumps is now consistent with the perfect delay convention. Since we can now account for the behaviour of the amoebae without having to postulate a double switch, we conclude that this hypothesis is superior to the previous one, i.e. that it is H/P, and not H, which is the correct critical variable.

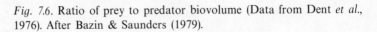

Fig. 7.6. Ratio of prey to predator biovolume (Data from Dent *et al.*, 1976). After Bazin & Saunders (1979).

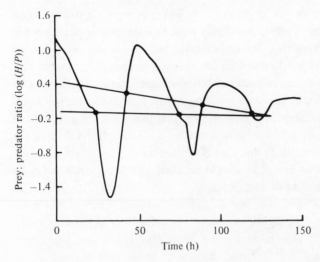

Fig. 7.7. Prey biovolume (Data from Owen, 1979). Jumps up and down in λ are distinguished by letters U and D, respectively, on the error bars. After Bazin & Saunders (1979).

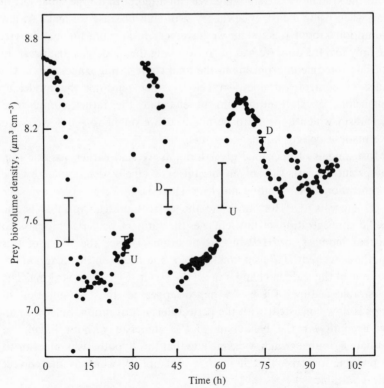

This example illustrates how catastrophe theory can enable us to obtain information from observations to which we are unable to fit a detailed model. Of particular importance, especially in the biological and social sciences, is the way in which we can make good use of noisy data. We can see from Figs. 7.4 and 7.6 that small errors in the values of H or H/P and in the locations of the peaks and troughs of P would leave the conclusions unaffected. Testing a model based on differential equations, on the other hand, would require values of the derivatives of all the variables, and these are notoriously sensitive to errors in the data.

The power of the method is demonstrated by Figs. 7.7 and 7.8. These are analogous to Figs. 7.4 and 7.6, except that they are from an experiment which was being conducted for a different purpose (Owen, 1979) and consequently fewer data relevant to our needs were collected. It is difficult to see how anything useful could be learned from an attempt to fit a model based on differential equations, yet even with a very large

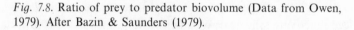

Fig. 7.8. Ratio of prey to predator biovolume (Data from Owen, 1979). After Bazin & Saunders (1979).

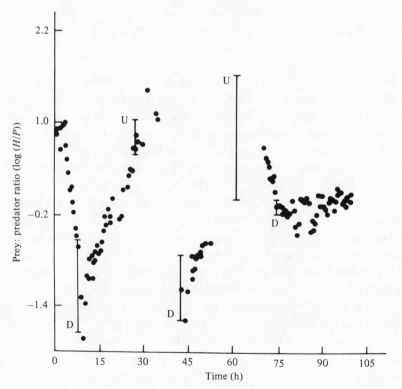

uncertainty in the locations of the jumps in λ the method based on catastrophe theory produces a result.

As is generally the case when we use catastrophe theory, we still do not know the mechanism by which the amoebal specific growth rate is controlled. We do, however, have a valuable clue; we are far more likely to succeed in finding the equations which govern the system if we use the correct independent variables. But what is more important is that we have obtained information which we can use directly. The sudden changes in specific growth rate indicate that there must be some sort of definite switch in the amoebae, and we would naturally like to know the identity of the signal to which they respond. Had the critical variable been H, then we would have been looking for any substance which is secreted by the bacteria, but if it is H/P then the substance must be one which is also modified by the amoebae. In fact at the same time that Bazin & Saunders (1978) pointed out that if folic acid was the signal then their result implied that the amoebae must be modifying it, Pan & Wurster (1978) independently reported that *Dictyostelium discoideum* do indeed inactivate folic acid. Whether or not folic acid does turn out to be the substance used by the amoebae to measure the relative prey density, this example illustrates how catastrophe theory can produce testable hypotheses.

In some respects this application is quite different from the first one, both in the arguments we used and in the conclusions we drew from them. The two results do, however, have a very important feature in common, and that is that they are both statements not about individual models but about classes of models. In the first example we established the existence of a large class of models with an important but unexpected property, while in the second example we were able to choose between two classes of models and, as it turned out, to reject the class which includes most of those usually used to describe systems of the given type. In both cases we have made some progress towards the construction of a model (in the usual sense) of the system. We have also gained knowledge of the system which we can use in further studies without our having to wait for the details of the mechanism to be elucidated. Phenomenological models, generally in the form of *ad hoc* equations, are sometimes also used for this purpose, but analyses based on catastrophe theory have the great advantage that the conclusions are far more robust, as they do not stand or fall with one particular choice of model.

8

Morphogenesis

One of the most interesting – and difficult – problems in biology is that of understanding development, the process by which a fertilized egg becomes first an embryo and then a fully formed organism. And an important aspect of development is morphogenesis, the creation of the various forms characteristic of the organism and its constituent parts. Of course the problem of form and the succession of forms is encountered in other branches of science as well, but it is in developmental biology that it is especially important. How is it that out of a single cell there can develop an organism which is to such a large extent the same as all others of the same species? How is it that this can happen even though different individuals within a species may be quite different in size and in certain details of their shape? And how is it that the process is so stable, allowing considerable variation in the environment and resisting many (though by no means all) perturbations?

In seeking an answer to these questions, the first step is to try to establish exactly what it is that the developmental process accomplishes. In the language of the earlier chapters of this book, what do we mean when we say that two individuals are 'of the same form'? It is clearly impossible to give a completely satisfactory definition, but the mathematical relation that most nearly captures the essential idea is that of topological equivalence. We can improve on this if we require that the homeomorphism must respect the types of tissue; this leaves us with a class of homeomorphisms under which a doughnut is not equivalent to a sphere even if we fill the hole with jam.

Now it is certainly true that even in this restricted sense no two humans are strictly homeomorphic, because we all have different numbers of pores in our skin and hairs on our heads. It is also true that topological equivalence, even if defined at a level of detail sufficiently low to avoid this objection, is not the whole story. Geometry also matters; roughly speaking all humans have similar proportions, and this distinguishes us from other primates who may be very close to us topologically. All the same, when compared with our variability in size and shape, humans, like other individuals within a species, do show a remarkable constancy in the number of bones, muscles, internal organs and other components, and in the connections between them. So while geometry cannot be ignored altogether, it seems that the control over the topology is stricter, and that it is the topology which is fundamental in development.

Confirmation of this idea can be found in D'Arcy Thompson's classic work *On Growth and Form* (D'Arcy Thompson, 1917). Anyone who has read this book – and anyone who has not is strongly advised to – is bound to have been impressed by the chapter on the 'theory of transformations', and especially by the famous illustrations, one set of which are reproduced here as Figs. 8.1–8.4. What D'Arcy Thompson discovered was that by drawing one species of fish on a rectangular coordinate grid, and then performing what we would call a simple diffeomorphism, he could obtain remarkably close likenesses of three different, though related, species. Similar transformations were found for a number of other examples, including the skulls of primates and other mammals, the pelvises of fossil birds and the carapaces of various crabs. This is strong evidence indeed for the claim that the topology is basic to the plan, with the geometry being filled in later and consequently being more susceptible to change during evolution.

But how is the topology specified? It is important to recognize that morphogenesis does not occur simply by cells of the correct type being formed in appropriate positions. It is true that later on a liver cell, for example, will divide to produce two liver cells, but in the early stages cells are not generally *determined* when they are formed; they are for some time capable of developing into one of a number of different final types. This can be demonstrated by grafting experiments in which a cell transferred from one part of the embryo to another develops like its new neighbours, and not like the cells in the region from which it came. (See e.g. Bellairs, 1971.)

Precisely how the fate of an individual cell is decided is not known, but it is clear that there are interactions between neighbouring cells and that there are physical and chemical gradients within the embryo. Many workers have used models based on these gradients to explain some of the various processes which occur during development. These models naturally differ in many respects, but they all assume that it is chiefly on account of these gradients that not all cells develop in the same way. (We recall that the genetic information in all cells other than gametes is identical.)

Let us accept this basic idea, and let us suppose further that the cells are also able to measure time, in the sense that there are time-dependent changes going on within them. And for the moment let us also suppose (as in Zeeman's account of frontier formation which we discussed in Chapter 7) that the fate of each cell is decided by a particular substance which we may call a morphogen and whose concentration is determined by the equilibrium states of a set of ordinary differential equations. The equilibrium states will themselves depend upon a number of parameters, but since there can only be four independent gradients (three in space and one in time) there are at most four independent control variables.

Fig. 8.1. *Polyprion.* From D'Arcy Thompson (1961).

Fig. 8.2. *Pseudopriacanthus altus.* From D'Arcy Thompson (1961).

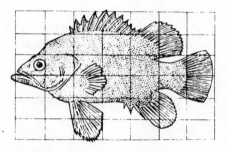

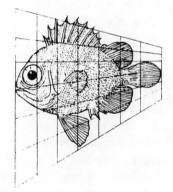

Fig. 8.3. *Scorpaena* sp. From D'Arcy Thompson (1961).

Fig. 8.4. *Antigonia capros.* From D'Arcy Thompson (1961).

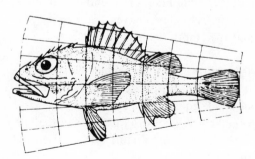

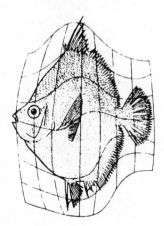

We are now on familiar ground. For the most part, the variation in the morphogen concentration, and consequently that in the observable properties of the cells, will be smooth. There will, however, be values of the four control variables for which different equilibria are possible, and this will result in the formation of well-defined frontiers between different types of tissue and between tissue and nothing (regions corresponding to empty regimes). It follows that in the neighbourhood of a point in space–time the shape of the structure being formed will be determined by the elementary catastrophes of codimension not greater than four.

As in our original outline of catastrophe theory, the conditions we have imposed are more restrictive than are really necessary. It can be shown, for example, that the singularities of a large class of partial differential equations (including the wave equation) propagate according to the elementary catastrophes (Guckenheimer, 1973). We have already seen an example of this, when we discussed the formation of caustics in terms of geometrical optics, which meant that we were working with the characteristics of the wave equation. Thom uses this idea in his book, couching many of his arguments in terms of 'wavefronts' and 'shock-waves', but he also writes of a more general approach, which he calls the 'metabolic model'. In this the equilibrium point of a system of ordinary differential equations is replaced by the more general concept of the attractor of a dynamic. For the present, however, we can profitably adopt the same tactic as before: to think in terms of a potential, while recognizing that our results are applicable to a much broader range of problems. Exactly how much broader is, to be sure, not yet known.

We now list the sorts of morphologies to which each of the seven elementary catastrophes can give rise; in this we follow Thom (1970). In most cases the Maxwell convention is appropriate, though with the proviso that it is only used to decide between finite attractors. Where, as is often the case, there is also a global minimum of potential at infinity, we shall suppose that some form of delay convention applies, for otherwise we obtain nothing but the empty regime.

Morphologies associated with the seven elementary catastrophes

The fold. We recall that the control space is the real line. To the left of the origin there is one stable equilibrium, to the right there are none. If we interpret the single control variable as spatial, then the fold represents a boundary; if we take the control variable as time then the fold represents beginning or ending.

The cusp. Depending on the choice of convention, the cusp represents a pleat or a fault (as in geology; see Fig. 8.5). If one of the control variables is time then it represents the action of separating (or uniting) or changing.

The swallowtail. We saw in Chapter 4 that for $u > 0$ the swallowtail divides the v–w plane into two regions, one with a single stable equilibrium, and one with none. For $u < 0$, however, a cusp catastrophe develops within the former region, causing a separation within it (Fig. 8.6). The swallowtail can therefore be interpreted as a split or furrow or, if u represents time, as the action of splitting or tearing.

The butterfly. The interesting case is naturally the one with the butterfly factor, t, negative, for otherwise we have nothing more than a cusp. If we then take the bias factor u as time, we obtain the sequence shown in Fig. 8.7. This is hard to observe directly because it is transient, but if it passes through a suitable material it can leave a cell joined to a vertical partition (Fig. 8.8). The butterfly can be interpreted spatially as a pocket, while in the temporal interpretation it corresponds to giving or receiving, or to the filling or emptying of a pocket.

The elliptic umbilic. There is only one stable equilibrium possible, and the boundary of the region in which it exists is shown in Fig. 8.9. If all the control variables are spatial, the elliptic umbilic represents a pointed structure such as a spike or a hair, whereas on the temporal

Fig. 8.5. A fault.

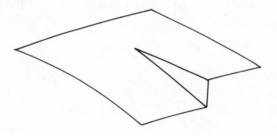

Fig. 8.6. The formation of a split.

Fig. 8.7. The formation of a pocket.

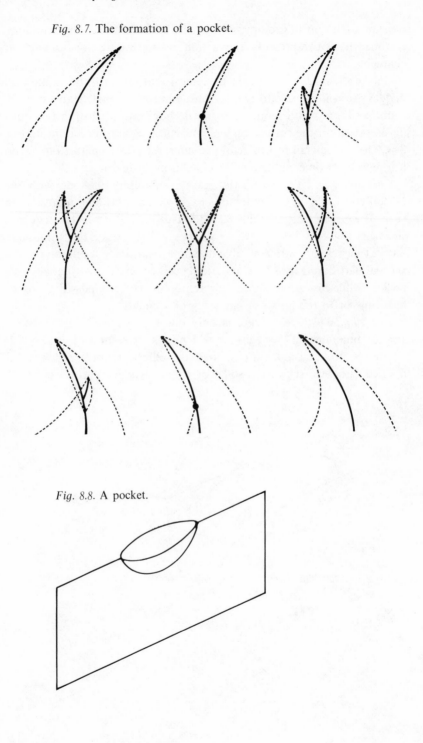

Fig. 8.8. A pocket.

interpretation it represents drilling or filling a hole. It can also represent a liquid jet, but not a stable hydrostatic configuration on account of the cusps; in biology it is possible for the shape to be stabilized by local processes.

The hyperbolic umbilic. Again there is only one stable equilibrium possible. A full sketch of the boundary of the region in which it exists is given in Fig. 4.9, and three cross sections are shown in Fig. 8.10. If we take *w* as time and interpret the sketches as successive cross sections of a wave, we see the crest of the wave becomes more and more pointed until it is actually angular (Fig. 8.10*b*), at which stage it breaks. There is, by the way, some disagreement as to whether or not this account gives a good description of the dramatic breaking of waves on a beach. The effect can, however, be clearly seen in 'symmetric breaking', which often occurs away from the shore.

The hyperbolic umbilic can also be interpreted as an arch, or, taking *w* as time, as collapsing or engulfing.

The parabolic umbilic. We did not give a complete description of the geometry of this catastrophe, and so we cannot provide a detailed account of its interpretations. On the other hand, the configurations shown in Fig. 4.14 do allow us to see how the parabolic umbilic can be interpreted spatially as (for example) a mushroom or a mouth and

Fig. 8.9. The boundary of the only non-empty regime of the elliptic umbilic.

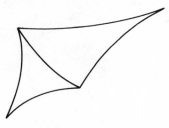

Fig. 8.10. Cross sections of the only non-empty regime of the hyperbolic umbilic: (*a*) $w < 0$, (*b*) $w = 0$, (*c*) $w > 0$.

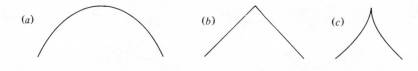

temporally as the opening or closing of a mouth or as piercing, ejecting or throwing.

To this set of catastrophes, which we may think of as the building blocks from which we may attempt to construct models of morpho-genesis, there are a few more shapes to be added. First, there is the simple minimum, x^2, which we may interpret as an object, or as being or lasting. Then there are the 'lips' and 'bec-à-bec' which we met in Chapter 4 as three-dimensional versions of the cusp. Thom also includes in his list the frontiers which separate two or more unrelated attractors, such as a line (in the case of two attractors) or the configurations shown in Fig. 8.11. The first of these, which arises from the conflict of three attractors, can be observed in cellular partitions and in Mach's reflexion in gas dynamics. Morphogenesis is thus seen as the result of conflict, either between separate attractors or between different regimes of the same attractor, and Thom (1972) underlines this with a quotation from Heraclitus: 'It should be known that war is universal, that strife is justice, and all things come into existence by strife and necessity.'

Applying the theory
It is one thing to see how catastrophe theory might have a contribution to make to the study of morphogenesis, but it is quite

Fig. *8.11.* Morphologies due to more than one organizing centre: (*a*) Conflict of three attractors, (*b*), (*c*) Possible transitions arising from the conflict of four attractors.

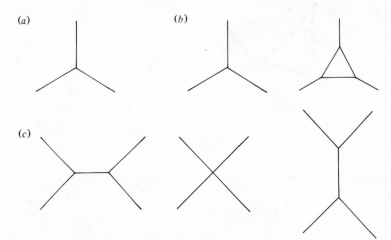

another actually to use it. Nothing that has been said in this chapter is to be taken as implying that the problem is all but solved, that all that remains is for developmental biologists to look carefully at embryos with lists of the seven elementary catastrophes beside their microscopes and all will be clear.

In the first place, elementary catastrophe theory is probably insufficient. Thom himself considers that much of what we observe can only be described by what he calls *generalized catastrophes*, and these are as yet not well understood. The essential idea of a generalized catastrophe is that an attractor which up to a certain time governs exclusively in a certain domain D then ceases to do so, and is replaced by a number of new attractors, each governing only part of D. What happens next is not easily predicted, but depending on the codimension of the new phases the domain can arrange itself into lumps (or, dually, bubbles) or into thin bands or filaments, the latter two being commonly encountered only in biology. A beautiful example of a generalized catastrophe can be seen – though only with the aid of high-speed photography – when a drop hits the surface of a fluid (Fig. 8.12). The resemblance between the shape thus formed and that of a hydroid polyp (Fig. 8.13) is also significant from our point of view.

Quite apart from the mathematical difficulties, however, development is a very complicated process, with many different activities going on at the same time, and with structures often arising through a combination of effects. We must not, for example, forget that cells can migrate during development. To give some idea of the sorts of arguments that will have to be used, we consider an account (Thom, 1973) of two different sequences of events which can follow a differentiation.

We begin with the separation of ectoderm from endoderm in a chick embryo. According to our list, separation can be modelled with a cusp catastrophe

$$V = x^4 + ux^2 + vx.$$

Initially, we may take u to be time and v to be the external–internal gradient, but we would expect that at a later stage the gradient will itself be affected by the reactions that are taking place so that we will no longer be able to consider it as a control variable. We then have to decide how to represent this change in the potential. The obvious way is simply to let v be a state variable, but we then find that the quadratic part of $V(x, v)$ is $ux^2 + vx$, and the value of the Hessian at the origin is -1 so there is no degeneracy. This is fine if we want a permanent entity, but it contributes nothing towards predicting the next change of form which will occur.

The next simplest modification of the potential is to write

$$v = \hat{v} + y,$$

where $\hat{v}$ is the externally determined contribution to the external–internal gradient and y is the locally determined contribution, but this too leads nowhere, for the same reason. So we try

$$v = \hat{v} + y^2,$$

Fig. 8.12. A drop hitting the surface of a fluid. From D'Arcy Thompson (1917).

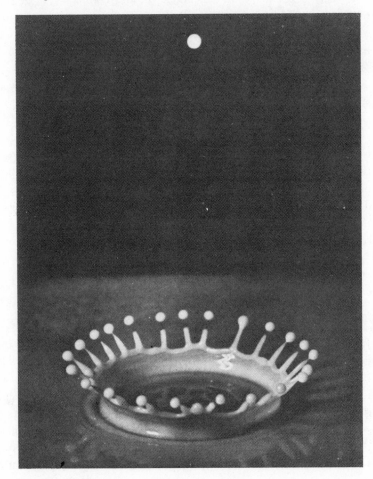

which gives

$$V(x, y) = x^4 + xy^2 + ux^2 + vx + sy^2 + ay,$$

where the last two terms have been added to stabilize what we recognize as the parabolic umbilic.

This potential is invariant under the transformation

$$y \mapsto -y, \quad a \mapsto -a.$$

Hence if a is the medio-lateral gradient there will be a bilateral symmetry in the embryo. Thus the reason that bilateral symmetry is so common in organisms may be that it is associated with what at an early stage in development is the simplest way forward. We complete the picture by taking the other new parameter, s, to vary slowly along the spinal axis, vanishing at Hensen's node. It causes the umbilic to be elliptic on the cephalic side and hyperbolic on the caudal side.

This account appears to give a good description of the process and at the very least it suggests that it might be profitable to study the separation of ectoderm and endoderm and the formation of the noto-chord together, rather than separately. But not only is it not necessarily the correct explanation of what happens; it is not even the only mor-phology which catastrophe theory predicts can arise out of a separation.

A second non-trivial way of incorporating v into the local metabolism is simply to let $v = y^2$. This implies a sudden change in the gradient and therefore seems less natural than the above, but it could happen if a local attractor $y = z = 0$ in the space of internal variables (of which x is originally the only essential variable) were to undergo a Hopf bifurcation which created a small attracting circle around the origin. The radius of

Fig. 8.13. (a), (b) More phases of a splash. (c) A hydroid polyp. From D'Arcy Thompson (1917).

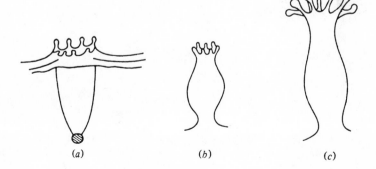

(a)   (b)   (c)

this limit cycle would in general depend on x, and we may denote it by $r(x)$.

In the y–z plane the trajectories would all spiral inwards or outwards towards the cycle. We eliminate the non-essential variable z in the usual way, by projecting onto the y-axis, and we find that there are now two stable critical points, at $\pm r(x)$. A local Liapounov function for the flow is

$$F = (y - r(x))^2 (y + r(x))^2.$$

The simplest choice for $r(x)$ which makes F a polynomial is $r(x) = \sqrt{x}$, and this leads us to the potential

$$V(x, y) = x^4 + ux^2 + y^4 + \lambda xy^2,$$

which is a double cusp. To stabilize this we need to add terms in x, y, xy, y^2, x^2y and x^2y^2, but this will give eight unfolding parameters altogether, and for a stable morphology in space–time the limit is four.

Let us suppose, however, that there is a bilateral symmetry in the system. Then we need only include even powers of y in the unfolding of V, and this brings the codimension down to five. But this is the algebraic codimension; the topological codimension, which is what really concerns us, is one less, i.e. four, and so the corresponding morphology can appear stably in space–time.

The above may all sound rather speculative, and even the recognition that the choice of potentials is dictated not by mere caprice or *ad hoc* rationalization but rather by considerations of simplicity and structural stability may leave the reader unconvinced. The only answer at this stage must be to wait and see; what we have seen here is a brief outline, a plan of attack on a very difficult problem. It would, however, be wrong to ignore morphogenesis altogether, even in a first encounter with catastrophe theory, for it was out of Thom's interest in this problem that catastrophe theory arose, and it may yet be in this field in which it will make its most significant contribution to science.

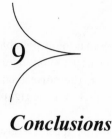

9

Conclusions

In this chapter we bring together the different ways in which we have seen that catastrophe theory can be applied, and we discuss how catastrophe theory compares with other methods as a means of explaining nature.

Applications of catastrophe theory

We began our study of the applications of catastrophe theory by quoting Thom's observation that they form a spectrum of different types. Now that we have seen a number of applications we can describe the spectrum in more detail.

At the extreme 'physical' end, catastrophe theory is used much like any other mathematical method, to help us discover the properties of a known, or at any rate postulated, dynamic. For example, when we are studying the buckling of elastic structures or the stability of ships (Zeeman, 1977*a*), the concept of a universal unfolding and the techniques for finding one can warn us when our analysis is structurally unstable and can be used to suggest what further effects are to be expected. Catastrophe theory can also be used to establish results which are true for a large class of systems, each with a known dynamic; Berry's (1976) work on caustics, which we discussed in Chapter 5, is a good example.

In the middle of the spectrum are the applications to the study of systems whose mechanisms are not known in sufficient detail for us to write down and solve the equations that describe them, but for which we are reasonably confident that we know the sorts of equations that are involved. This is often the case in biology, because we generally suppose that the system can be discussed in terms of ordinary differential equations (which govern the biochemical reactions) and, what is more, that a steady-state approximation is appropriate. In these applications, catastrophe theory can enable us to derive results without the necessity of postulating – and justifying – a particular mechanism. It can also assist in the problem of elucidating the mechanism by making it possible for us to tackle the problem in stages. It can be much easier to eliminate several classes of possible dynamics first and then seek the correct equations from among those that remain, than to have to pick out the right answer in one fell swoop.

The applications at the 'metaphysical' end of the spectrum are, in the first instance, metaphors. It is, of course, sometimes difficult to convince anyone else of the value of a metaphor as it is chiefly heuristic; once we have grasped the point we no longer need it. All the same, sketching a cusp does generally make it easier to understand and discuss the five typical features of the cusp catastrophe and to see them as different aspects of the same thing. The interest shown by some social scientists in catastrophe theory is an indication of the value to them of a conceptual framework which goes beyond linearity and which recognizes that discontinuous effects need not have discontinuous causes.

It is, however, important to remember that what catastrophe theory provides are never *just* metaphors. The surfaces are not sketched at random, nor are they *ad hoc* constructions to fit particular situations. On the contrary, they are chosen on grounds of simplicity and structural stability. We cannot be certain that catastrophe theory is applicable in any individual case, but it is a useful working hypothesis. Moreover, if we do succeed in fitting a model based on a catastrophe, then it is natural to ask what it is about the system that makes this possible, and this too can lead to further advances.

In this brief account of the range of applications of catastrophe theory we do not claim to have given all the different ways in which it can be used. Indeed, even if we had set out to include all those known at the time of writing, new ones would be bound to appear. And we have not attempted to describe how Thom himself has used catastrophe theory as the basis for some highly original speculation in a number of fields (see Thom, 1972, for examples) though perhaps those who are capable of this sort of thought need no explicit invitation. It will be clear, however, that it is impossible to summarize in a sentence or two what catastrophe theory is and how to use it. The best we can do is probably to remark that there is a feature common to all but the most 'physical' of the applications, and that is the use of Ockham's razor. When we do not know what the mechanism of a system is, we make the assumption that it is the simplest that is consistent with the observations, and then see what else this implies about the system. The role of catastrophe theory is to tell us, in a well-defined way, what the 'simplest sort' of mechanism is.

Catastrophe theory and explanation

We have seen that catastrophe theory can be successfully applied in both the 'hard' and the 'soft' sciences. The reader may, however, still feel somewhat uncomfortable with the theory, on the grounds that it often seems to provide only a description of the system, not an explanation, and certainly not a mechanism. As this issue is probably the source of much of the controversy that has surrounded catastrophe theory, it is worth considering it.

In the first place, it is by no means clear what we really mean when we speak of an 'explanation'. As long ago as the fourth century B.C., Aristotle observed that one could distinguish four quite different sorts of causes of a thing or an event, and that what constitutes an explanation depends rather on what one happens to be interested in. And nowadays most scientists would accept that we can never have an ultimate explanation for anything, and that none of our theories can accomplish more than to arrange our knowledge of nature into an orderly system, with more and more links being established between previously unconnected phenomena. Thus when Thom (1975) describes the aim of catastrophe theory as 'reducing the arbitrariness of description', he is not setting it apart from the rest of science, for he is expressing the same point of view as Planck (1925) who wrote 'As long as Natural Philosophy exists, its ultimate aim will be the correlating of various physical observations into a unified system, and, where possible, into a single formula.'

A full discussion of the nature of knowledge and explanation in science is clearly beyond the scope of this book, but we must not forget that there are fundamental unsolved – and perhaps unsolvable – problems involved. We may be more aware of them in catastrophe theory because it is so new that we are still at the stage of deriving the techniques for using it, but we can see the same sorts of difficulties if we look closely enough at any science or even, as Lakatos (1976) argued in his thought-provoking book *Proofs and Refutations*, mathematics itself. It is unreasonable to criticize catastrophe theory for failing to accomplish what no theory can.

As for mechanisms, it is true enough that models in physics are often mechanistic, that is, they are translations into mathematical terms of the known or supposed physical causes of the phenomena to be explained. But this is not always the case. It is not true of, say, the Schrödinger equation of quantum mechanics, or the field equations of general relativity. In fact, both theories were strongly attacked on precisely these grounds (see, for example, Stark, 1938) although they have by now achieved general acceptance. So if we are reluctant to work with models which postulate no definite mechanism, we must reject more than just catastrophe theory. Alternatively, if we are reconciled to the modern view of science, then we may ask with Whitehead (1926) what sense there is in talking about a mechanical explanation when we do not know what we mean by mechanics.

Yet even if the laws of physics are not the direct insight into the workings of the universe that they were once thought to be, they do represent hypotheses which have been verified in many different circumstances. We therefore feel entitled to take them as axioms for the derivation of further predictions about the behaviour of physical systems, at least until they are found to be inadequate and are superseded by other axioms. And when we can construct from these axioms a model which we believe to be a good mathematical representation of the mechanism of the system, and when the predictions of the model are in good quantitative agreement with the observations, we may well feel that there is a real sense in which we have an 'explanation' of the phenomenon. Catastrophe theory does not usually provide the same sort of explanation, and so in this respect the traditional approach has the edge – when it can be made to work.

As we remarked in the introduction, in physics the standard methods generally work well. It is therefore likely that catastrophe theory will play a relatively less important, though useful, role in physics. But the situation is quite different in the biological and social sciences. Here we do not have well-established laws and precise quantitative observations. It is pointless to argue that a detailed mechanistic model would provide a better explanation than catastrophe theory can, when no such model is available, or likely to be.

It is, however, very easy to be misled in this respect by some of the work which has been done in, for example, theoretical biology. Consider the Lotka–Volterra equations, which we met in Chapter 7:

$$\dot{H} = \alpha H - \mu HP,$$
$$\dot{P} = \nu HP - \beta P.$$

These equations may look like the sort of thing we are accustomed to in physics, but in fact they are really quite different. They are not the 'Newton's laws' of theoretical ecology; at best they are a very crude approximation to the interactions that are actually taking place. And while we can obtain quantitative solutions of these equations if we wish, this will not give us quantitative results concerning the systems they purport to represent. Nowhere do we find tables of the constants of the Lotka–Volterra equations (or even modified versions of them) which would enable us to make the same sort of confident predictions that, say, Snell's law and a list of refractive indices permit.

The most we can hope to gain from these equations is some understanding of the general behaviour of the system. Volterra's own aim, after all, was not to forecast the numbers of fish which would be caught in the Adriatic in any given year, but to demonstrate that the oscillations arose out of the predator–prey interaction alone. As it happens, Volterra may well have been wrong, but his work is an early, and probably still the best-known, example of a technique which has proved quite useful in theoretical biology. Faced with a system which is too complex to be modelled accurately, we construct a mathematical system which we consider has certain essential properties in common with it. We then analyse the model and draw conclusions which we hope will be applicable to the real system as well. It is, of course possible that our results are artefactual, but so long as we are aware of this danger we can generally think of ways of testing them. Interesting examples of this approach include the work of Goodwin (1963) on oscillators, and of Kaufmann (1969) on binary nets. There are also some useful results in ecology (see Maynard Smith, 1974) though there is a tendency for workers in that field to treat their equations with more respect than they really deserve.

Thus when we look carefully at how mathematics is used in biology, we discover that the applications of catastrophe theory are not quite so different in principle as they appear at first glance. Moreover, since catastrophe theory was designed to be used in this way, it is better suited to it.

Most other methods require us to construct some sort of model of the system. This is a natural approach in physics, where, as we have already mentioned, the model constitutes a form of explanation in itself. In the sort of application we are concerned with here, on the other hand, it implies that between the hypotheses with which we begin and the conclusions with which we hope to end, we have to insert additional hypotheses not because we believe them to be true or even because we want to test them, but simply in order to proceed. When we use catastrophe theory we do not generally have to do this. This has the obvious advantage that we can be more confident that our conclusions follow from our original hypotheses, and are not consequences of the extra ones. And, as we saw in Chapter 7, there are cases in which catastrophe theory can produce results when it is hard to see how any plausible model could be found.

Another advantage of catastrophe theory is that conclusions which are based on it are guaranteed to be structurally stable, which is not always the case with other methods. For example, the Lotka–Volterra equations are structurally unstable; the solutions of almost any pair of equations 'near' them are not persistent oscillations. It is therefore unlikely that where persistent oscillations are observed in nature they are due to an interaction of the type implied by these equations. Of course most of the models used in ecology and elsewhere are structurally stable, but there is always the chance that they are not, and we know from the earlier chapters of this book that this is particularly likely when in trying to study a complex system we allow considerations of mathematical tractability to tempt us into using a model with too few parameters.

Catastrophe theory possesses these advantages because it is a topological theory, and therefore produces qualitative results directly. It is not, of course, the only theory with this property, any more than it is the only theory which deals with discontinuities. There is a considerable, and rapidly growing, literature available under the general headings of topological dynamics and bifurcation theory. But almost all the mathematics used by biologists is of the kind that produces quantitative results about special cases, which have to be converted. In biology, Rutherford's famous dictum can often be reversed: there are circumstances in which quantitative is just bad qualitative.

When catastrophe theory is applied in physics, it is reasonable to judge it by the standards of the other techniques available to physicists, and, as we have seen, it is indeed capable of producing the same sorts of results. But when catastrophe theory is used in biology, it – like any other method used in that subject – must be judged not by how closely it follows the paradigm of theoretical physics but by how much it contributes to our understanding of biological phenomena.

There is, of course, another side to the coin. If we believe that the results we may expect in much of theoretical biology are likely to be different from those which are usual in physics, then we are not entitled to make the same claims for them. We have already acknowledged that they are generally more tentative; in particular, if we have used catastrophe theory then there is usually a crucial assumption of simplicity at some point in the argument. Moreover, if a mechanistic model is found to be in good agreement with the observations, then we have immediately achieved something, in that we have evidence in support of our hypotheses concerning the mechanism. Merely fitting a cusp to some data, on the other hand, accomplishes comparatively little by itself. What matters is what happens next. Simple examples such as Zeeman's analysis of the behaviour of dogs are indeed useful, but more as illustrations of catastrophe theory than as contributions to (in this case) ethology. There are now enough illustrations, and in applying catastrophe theory in biology and the social sciences we may claim success not when we have found a catastrophe on Thom's list that matches our observations, but when by doing this (or even, as we have seen, by failing to do this) we have learned something new about the system we are studying.

Exercises

1. Find the potential energy of a gravitational catastrophe machine which is like the one described in Chapter 1 except that its perimeter is the ellipse $b^2x^2 + a^2y^2 = a^2b^2$. Show that the sudden jumps in position occur as the magnet crosses the perimeter of a curvilinear diamond whose vertices lie on the coordinate axes. Show that there are ordinary cusps at two of the vertices and dual cusps at the other two. Comment on the case $a = b$.

Hence predict how the machine will respond as the magnet is moved along different paths. How would the behaviour be affected if the perimeter of the machine was a smooth curve which was only approximately an ellipse? Describe also the response of a circular machine which is (*a*) perfectly constructed (*b*) slightly imperfectly constructed. (If you build an elliptical catastrophe machine to check your answer, try it out on an inclined plane as well. See Poston & Stewart (1976) for an analysis.)

2. Derive the canonical forms and universal unfoldings of all the elementary catastrophes of codimension 5.

3. Prove that the cross sections of the bifurcation set of the butterfly catastrophe for $t \geq 0$ are as shown in Fig. 4.11.

4. Consider the canonical wigwam catastrophe
$$x^7 + sx^5 + tx^4 + ux^3 + vx^2 + wx.$$
Let B^* be the section of the bifurcation set by a surface $s = $ constant, $t = $ constant, $u = $ constant; B^* is then a curve in a plane parallel to the vw plane.
(i) Show that B^* passes through the origin of the plane, and that the tangent to the origin is in the v-direction. Show that this is the only horizontal or vertical tangent.
(ii) Identify t as the bias factor.
(iii) Taking $t = 0$, sketch the analogues of Figs 4.13 and 4.14. (Remember that $v' = w' = 0$ is a necessary but not sufficient condition for a cusp.)

5. Let B^* be the analogous curve for the general cuspoid
$$x^n + ax^{n-2} + bx^{n-3} + \ldots + ux^3 + vx^2 + wx,$$
and suppose $n \geq 5$.

(i) As part (i) of question 4.

(ii) For what values of n is it possible to pick out a bias factor?

(iii) Sketch B^* for the reduced cuspoids

(a) $x^n + ax^{n-2} + vx^2 + wx,$

(b) $x^n + ux^3 + vx^2 + wx.$

Note that the cases n odd and n even must be considered separately.

6. Equations which are sometimes used to represent the behaviour of a single species of microorganism growing in a chemostat are
$$\dot{x} = (\mu - D)x,$$
$$\mu = \mu_m K_i s/(s^2 + K_i s + K_i K_s),$$
$$\dot{s} = D(s_r - s) - \mu x/Y.$$

Here x is the biomass, μ is the specific growth rate, s is the nutrient concentration, D is the dilution rate, and the other quantities are all constants.

Suppose the system is allowed to come to equilibrium and the dilution rate is then slowly increased. Interpret the subsequent behaviour as a fold catastrophe. Find the diffeomorphism which relates the actual state and control variables to the x and u of the canonical fold.

7. Describe the envelope of the family of straight lines given by
$$x + \lambda y = 2\lambda^3 - \lambda^5,$$
where λ is a continuously varying parameter.

8. By determining the stability of the solutions for the case $a = \omega = 0$ (or otherwise), show that the Duffing equation
$$\ddot{x} + k\dot{x} + x + ax^3 = F \cos(1 + \omega)t.$$
leads to a pair of cusp catastrophes if $k > 0$, but a pair of dual cusps if $k < 0$. Describe the behaviour of a system which is governed by Duffing's equation with negative damping. Comment on the possible application to brain modelling.

9. Show that, within a certain range of the control variables, examples in the social sciences which can be fitted to a cusp catastrophe can also be fitted to a dual butterfly. What different predictions does the latter model make? For any one particular model, suggest interpretations for the bias and butterfly factors.

10. A variable Q was measured as a function of time, and the result is part (a) of the figure below. Parts (b) and (c) are plots of two variables X and Y. In the absence of any additional information, which of X and Y is more likely to be governing Q?

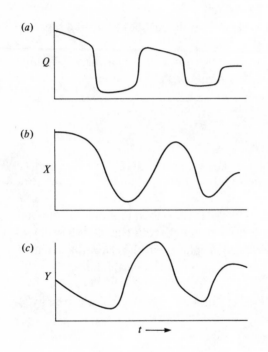

Appendix

Elementary catastrophes of codimension ≤ 5

Fold: $x^3 + ux$
Cusp: $x^4 + ux^2 + vx$
Swallowtail: $x^5 + ux^3 + vx^2 + wx$
Butterfly: $x^6 + tx^4 + ux^3 + vx^2 + wx$
Wigwam: $x^7 + sx^5 + tx^4 + ux^3 + vx^2 + wx$
Elliptic umbilic: $x^3 - xy^2 + w(x^2 + y^2) + ux + vy$
Hyperbolic umbilic: $x^3 + y^3 + wxy + ux + vy$
Parabolic umbilic: $y^4 + x^2y + wx^2 + ty^2 + ux + vy$
Symbolic umbilic: $x^3 + y^4 + sxy^2 + ty^2 + uxy + vy + wx$
Second elliptic umbilic: $x^2y - y^5 + sy^3 + ty^2 + ux^2 + vy + wx$
Second hyperbolic umbilic: $x^2y + y^5 + sy^3 + ty^2 + ux^2 + vy + wx$

References

Bazin, M. J. & Saunders, P. T. (1978). Determination of critical variables in a microbial predator–prey system by catastrophe theory. *Nature, London,* **275**, 52–4.

Bazin, M. J. & Saunders, P. T. (1979). An application of catastrophe theory to the study of a switch in *Dictyostelium discoideum.* In *Kinetic Logic – a Boolean Approach to the Analysis of Complex Regulatory Systems,* ed. R. Thomas. Berlin: Springer.

Bellairs, R. (1971). *Developmental Processes in Higher Vertebrates.* London: Logos Press.

Berkley, G. (1734). *The Analyst.* London: J. Tonson. (Reprinted in *The Works of George Berkley, Bishop of Cloyne,* ed. A. A. Luce & T. E. Jessop, vol. 4, pp. 65–102. London: Nelson, 1951.)

Berry, M. V. (1976). Waves and Thom's theorem. *Advances in Physics,* **25**, 1–26.

Bröcker, Th. & Lander, L. (1975). *Differentiable Germs and Catastrophes.* Cambridge University Press.

Cooke, J. & Zeeman, E. C. (1976). A clock and wavefront model for the control of the number of repeated structures during animal morphogenesis. *Journal of Theoretical Biology,* **58**, 455–76.

Dent, V. E., Bazin, M. J. & Saunders, P. T. (1976). Behaviour of *Dictyostelium discoideum* amoebae and *Escherichia coli* grown together in chemostat culture. *Archives of Microbiology,* **109**, 187–94.

Elsdale, T., Pearson, M. & Whitehead, M. (1976). Abnormalities in somite segmentation induced by heat shocks to *Xenopus* embryo. *Journal of Embryology and Experimental Morphology,* **35**, 625–35.

Fisher, G. H. (1967). Preparation of ambiguous stimulus materials. *Perception and Psychophysics,* **2**, 421–2.

Fowler, D. (1972). The Riemann–Hugoniot catastrophe and van der Waals equation. In *Towards a Theoretical Biology* 4, *Essays,* ed. C. H. Waddington, pp. 1–7. Edinburgh University Press.

Godwin, A. N. (1971). Three dimensional pictures for Thom's parabolic umbilic. *Publications mathématiques. Institut des hautes études scientifiques, Paris,* **40**, 117–38.

Goodwin, B. C. (1963). *Temporal Organization in Cells.* London: Academic Press.

Guckenheimer, J. (1973). Catastrophes and partial differential equations. *Annales de l'Institut Fourier, Université de Grenoble*, **23**, 31–59.

Haken, H. (1977). *Synergetics – An Introduction*. Berlin: Springer.

Hilton, P. J. (ed.) (1976). *Structural Stability, the Theory of Catastrophes and Applications in the Sciences*. Berlin: Springer.

Holmes, P. J. & Rand, D. A. (1976). The bifurcations of Duffing's equation: An application of catastrophe theory. *Journal of Sound and Vibration*, **44**, 237–53.

Isnard, C. A. & Zeeman, E. C. (1976). Some models from catastrophe theory in the social sciences. In *The Use of Models in the Social Sciences*, ed. L. Collins, pp. 44–100. London: Tavistock Publications.

Kaufmann, S. A. (1969). Metabolic stability and epigenesis in randomly constructed genetic nets. *Journal of Theoretical Biology*, **22**, 437–67.

Lakatos, I. (1976). *Proofs and Refutations*. Cambridge University Press.

Lorenz, K. (1966). *On Aggression*. London: Methuen.

Lu, Y.-C. (1976). *Singularity Theory and an Introduction to Catastrophe Theory*. Berlin: Springer.

Maynard Smith, J. (1974). *Models in Ecology*. Cambridge University Press.

Owen, B. A. (1979). Ph.D. Thesis, London University.

Pan, P. & Wurster, B. (1978). Inactivation of the chemoattractant folic acid by cellular slime molds and identification of the reaction product. *Journal of Bacteriology*, **136**, 955–9.

Planck, M. (1925). *A Survey of Physics* (translated by R. Jones & D. H. Williams). London: Methuen.

Poston, T. (1976). Various catastrophe machines. In Hilton (1976, pp. 111–26).

Poston, T. & Stewart, I. N. (1976). *Taylor Expansions and Catastrophes*. London: Pitman.

Poston, T. & Stewart, I. N. (1978a). *Catastrophe Theory and its Applications*. London: Pitman.

Poston, T. & Stewart, I. N. (1978b). Nonlinear modelling of multistable perception. *Behavioural Science*, **23**, 318–34.

Poston, T. & Woodcock, A. E. R. (1973). On Zeeman's catastrophe machine. *Proceedings of the Cambridge Philosophical Society*, **74**, 217–26.

Stark, J. (1938). The pragmatic and the dogmatic spirit in physics. *Nature, London*, **141**, 770–1.

Thom, R. (1970). Topological models in biology. In *Towards a Theoretical Biology* 3. *Drafts*, ed. C. H. Waddington, pp. 89–116. Edinburgh University Press.

Thom, R. (1972). *Stabilité Structurelle et Morphogénèse*. Reading, Mass.: Benjamin. (English translation by D. H. Fowler, 1975: *Structural Stability and Morphogenesis*. Reading, Mass.: Benjamin.)

Thom, R. (1973). A global dynamical scheme for vertebrate embryology. In *some Mathematical Questions in Biology IV. Lectures on Mathematics in the Life Sciences*, vol. 5, ed. J. D. Cowan, pp. 3–45. Providence: American Mathematical Society.

Thom, R. (1975). Answer to Christopher Zeeman's reply. In *Dynamical Systems – Warwick* 1974, ed. A. Manning, pp. 384–9. Berlin: Springer.

Thom, R. (1976). The two-fold way of catastrophe theory. In Hilton (1976, pp. 235–52).

Thompson, D'A. W. (1917). *On Growth and Form.* Cambridge University Press. (Abridged edition, ed. J. T. Bonner, 1961. Cambridge University Press.)

Thompson, J. M. T. & Hunt, G. W. (1973). *A General Theory of Elastic Stability.* London: Wiley.

Trotman, D. J. A. & Zeeman, E. C. (1976). The classification of elementary catastrophes of codimension ≦5. In Hilton (1976, pp. 263–327).

Wasserman, G. (1976). (r, s)-stable unfoldings and catastrophe theory. In Hilton (1976, pp. 253–62).

Whitehead, A. N. (1926). *Science and the Modern World.* Cambridge University Press.

Zeeman, E. C. (1972a). A catastrophe machine. In *Towards a Theoretical Biology 4. Essays,* ed. C. H. Waddington, pp. 276–82. Edinburgh University Press.

Zeeman, E. C. (1972b). Differential Equations for the heartbeat and nerve impulse. In *Towards a Theoretical Biology 4. Essays,* ed. C. H. Waddington, pp. 8–67. Edinburgh University Press.

Zeeman, E. C. (1974). Primary and secondary waves in developmental biology. In *Some Mathematical Questions in Biology VIII. Lectures in Mathematics in the Life Sciences,* vol. 7, ed S. A. Levin, pp. 69–161. Providence: American Mathematical Society.

Zeeman, E. C. (1976a). Catastrophe Theory. *Scientific American,* **234** (part 4) 65–83. (An expanded version of this paper appears in (Zeeman (1977b).)

Zeeman, E. C. (1976b). The umbilic bracelet and the double cusp catastrophe. In Hilton (1976, pp. 328–66).

Zeeman, E. C. (1976c). Euler Buckling. In Hilton (1976, pp. 373–95).

Zeeman, E. C. (1976d). Duffing's equation in brain modelling. *Bulletin of the Institute of Mathematics and its Applications,* **12,** 207–14.

Zeeman, E. C. (1976e). Brain Modelling. In Hilton (1976, pp. 367–72).

Zeeman, E. C. (1977a). A catastrophe model for the stability of ships. In *Geometry and Topology, Rio de Janeiro, 1976,* ed. J. Palis & M. do Carno, pp. 775–827. Berlin: Springer.

Zeeman, E. C. (1977b). *Catastrophe Theory.* Reading, Mass.: Addison-Wesley. (Consists of reprints of Zeeman's papers. including all those cited here.)

Zeeman, E. C. (1978). A dialogue between a mathematician and a biologist. *Biosciences Communications,* **4,** 225–40.

Author index

141

Subject index